牛津少儿英汉动物百科

Oxford English-Chinese Visual Dictionary of Animals
the dictionary that shows you what words mean

张劲硕 译

2017年·北京

English text and illustrations originally published as Oxford Visual Dictionary of Animals by Oxford University Press, Great Clarendon Street, Oxford © Oxford University Press 2015

英语原版 Oxford Visual Dictionary of Animals 由牛津大学出版社出版

Great Clarendon Street, Oxford © Oxford University Press 2015

This English-Chinese edition published by The Commercial Press by arrangement with Oxford University Press (China) Ltd for distribution in the mainland of China only and not for export therefrom

本书由商务印书馆和牛津大学出版社（中国）有限公司共同出版，

仅在中国大陆地区发行，不得出口到其他地区。

Text Copyright © Oxford University Press (China) Ltd and The Commercial Press 2017

©牛津大学出版社（中国）有限公司，商务印书馆，2017

Oxford is a registered trademark of Oxford University Press

Oxford 是牛津大学出版社的注册商标

The Commercial Press has made some changes to the original work in order to make this edition more appropriate for readers in the mainland of China.

商务印书馆对原书进行了个别修改，使其更符合中国读者的需要。

出版说明

商务印书馆与牛津大学出版社的合作由来已久,"牛津进阶"系列词典一直深受我国英语学习者欢迎。近年来更是加大了英语学习类图书的出版力度,推出了《牛津小学生英汉双解词典》等图书。此次出版《牛津少儿英汉动物百科》(Oxford English-Chinese Visual Dictionary of Animals)是我馆少儿英语类产品开发的又一成果。

《牛津少儿英汉动物百科》英文版是牛津大学出版社为少年儿童编写的彩色动物百科,自2015年在英国出版以来,深受国外小读者和家长喜爱。本书向读者介绍世界各地的野生动物,全书结构清晰合理,内容兼顾知识和趣味,英语简洁地道,译文准确易懂,插图色彩鲜艳、栩栩如生,不仅能让小读者学到动物知识,还能帮助他们学习英语、扩大词汇量。

商务印书馆从牛津大学出版社引进本书,旨在为我国少儿读者提供一本优秀的英汉对照课外读物。我馆聘请中国科学院动物研究所张劲硕博士担任双语版的汉语翻译和审订工作,保证译文科学准确。双语版不仅增加与英文对应的汉语译文,还为小读者提供难读汉字的汉语拼音,并辅以全书英文的朗读音频和点读功能,书后的索引方便读者从英汉双语入手查询对应的内容。衷心希望本书能受到小读者的喜爱,让他们通过本书了解动物,走近自然,并学到地道的英语。

商务印书馆编辑部
2017年1月

本书二维码音频收听使用方法:

每章音频收听

本书章节标题页提供"本章音频"的二维码,使用手机扫描此二维码,即可收听对应章节的朗读录音。

Contents 目录

使用说明	4-7
Animal life 动物总览	8-9
Animal behaviour 动物行为	10-11
All kinds of animals 各类动物	**12-25**
Mammals 哺乳动物	14-15
Birds 鸟类	16-17
Reptiles 爬行动物	18-19
Amphibians 两栖动物	20-21
Fish and other sea creatures 鱼类和其他海洋动物	22-23
Insects and minibeasts 昆虫和其他无脊椎动物	24-25
Rainforest creatures 雨林动物	**26-35**
Rainforest habitats 雨林栖息地	28-29
Life on the rainforest floor 林地层动物	30-31
Creatures of the canopy 树冠层动物	32-33
In the Amazon rainforest 在亚马孙雨林	34-35
Forest and woodland wildlife 森林和林地动物	**36-45**
Forest habitats 森林栖息地	38-39
Deciduous forest creatures 落叶林动物	40-41
Life on a woodland floor 林地地面动物	42-43
Evergreen forest wildlife 常绿林动物	44-45
Mountain animals 山地动物	**46-51**
Mountain habitats 山地栖息地	48-49
Life in the mountains 山地动物	50-51
Grassland wildlife 草原动物	**52-63**
Grassland habitats 草原栖息地	54-55
On the savannah 在稀树草原上	56-57
More savannah creatures 稀树草原上的另一些动物	58-59
Grassland and moorland wildlife 草原和高沼地动物	60-61
In the bush 在荒野中	62-63

Contents 目录

River, lake and wetland wildlife 河流、湖泊和湿地的动物 — 64-75

- Water habitats 水域栖息地 — 66-67
- River creatures 河流动物 — 68-69
- Life on the river 河上的动物 — 70-71
- Lake and pond wildlife 湖泊和池塘动物 — 72-73
- Wetland animals 湿地动物 — 74-75

Desert wildlife 荒漠动物 — 76-81

- Desert habitats 荒漠栖息地 — 78-79
- Desert creatures 荒漠动物 — 80-81

Life in the ocean 海洋动物 — 82-95

- Oceans 海洋 — 84-85
- On the seashore 在海滩上 — 86-87
- In the ocean 在海洋中 — 88-89
- More ocean life 另一些海洋动物 — 90-91
- Coral reefs 珊瑚礁 — 92-93
- Creatures of the deep 深海动物 — 94-95

Animals of the polar regions 极地动物 — 96-103

- Polar habitats 极地栖息地 — 98-99
- Arctic and Antarctic wildlife 北极和南极的动物 — 100-101
- Life in the tundra 苔原动物 — 102-103

- Widespread creatures 分布广泛的动物 — 104-105
- Widespread birds 分布广泛的鸟类 — 106-107

Vocabulary builder 扩展词汇 — 108-115

- Animal words 动物词汇 — 108-111
- Animal word origins 动物词语的词源 — 112-114
- Animal idioms 动物习语 — 115

- Animal detective quiz 动物知识小测验 — 116-117
- English-Chinese index 英汉索引 — 118-122
- Chinese-English index 汉英索引 — 123-130
- Quiz answers 小测验答案 — 131-132

使用说明

这本书里不仅有许多动物词汇，还有丰富的知识，让你能够一边探索自然界一边学单词！本书分为九部分，每部分用不同颜色区分，有引人入胜的开篇场景画和主题概述。每页顶部标明了每部分的主题（如雨林动物、极地动物），页边注明了每页的具体话题。

如何使用本书？

本书通过图画、场景和示意图来介绍单词，让你轻松地查到想要的单词，并在寻找的过程中学到更多。

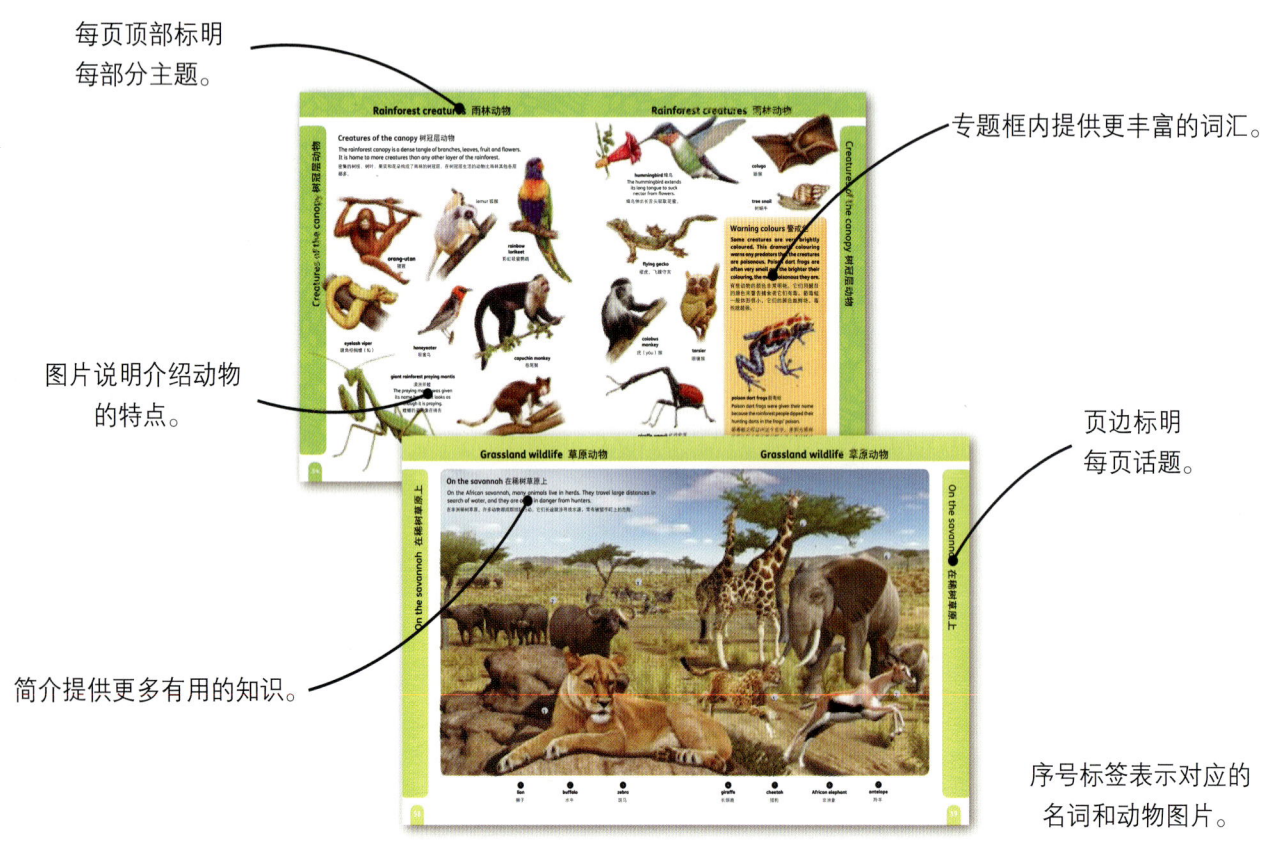

- 每页顶部标明每部分主题。
- 图片说明介绍动物的特点。
- 简介提供更多有用的知识。
- 专题框内提供更丰富的词汇。
- 页边标明每页话题。
- 序号标签表示对应的名词和动物图片。

如何查找单词？

查词有好几种方法。你可以在目录中查找主题，或者用每部分的颜色查找，还可以在书后的索引中查找。书中有几百个动物名称，还有与动物生活相关的其他词汇。

使用说明

本书的结构

本书的开头是对动物和动物行为的概述，接着介绍主要的动物类群。之后分八部分描述各类栖息地和在那里生活的动物。

"分布广泛的动物"这部分介绍在世界各地都能见到的动物。书后的"扩展词汇"列出了更多在谈论动物时可用到的词汇。

第8-11页

动物总览和行为

这部分介绍地球上千差万别的各种动物，研究不同动物如何移动、进食及运用它们的感官。还介绍动物的行为，如求偶、迁徙、冬眠。本部分的描述包括世界各地的动物。

第12-25页

各种动物

动物的主要类群有六类：
- mammals 哺乳动物
- amphibians 两栖动物
- birds 鸟类
- fish and other sea creatures 鱼类和其他海洋动物
- reptiles 爬行动物
- insects and minibeasts 昆虫和其他无脊椎动物

在这一部分，你能学到每类动物的特征。配有标签的图画帮你了解动物身体的各个部位。专题框介绍动物的发育过程和行为。

5

使用说明

第26-103页
动物栖息地

接下来你会认识八类不同的栖息地，或者说八类生物群系。

这八类栖息地分别是：

- rainforests 雨林
- forests and woodlands 森林和林地
- polar regions 极地
- grasslands 草原
- rivers, lakes and wetlands 河流、湖泊和湿地
- mountains 山地
- ocean zones 海洋
- deserts 荒漠

介绍每类栖息地时，首先都会展示一幅精美的场景画，带你走进新的环境。场景画之后有一段对栖息地和动物的介绍。每类栖息地在世界的分布情况，都在地图上标示得清清楚楚，各类栖息地不同的生存环境也都有讲解。图片旁的说明文字介绍动物的特点和行为。

在迥然不同的各类栖息地中，有新奇或熟悉的兽类、鸟类和昆虫等待你去发现。

第104-107页
分布广泛的动物

有些昆虫、鸟类等动物在全世界各类栖息地都有。第104—107页介绍的动物在世界任何地方都可能见到。

使用说明

扩展词汇

第108-111页
动物词汇

这部分是与动物有关的词汇。有动物幼崽的词汇列表、动物群体的说法，还有表示动物叫声的单词。快去看看这几页，学几个有趣的单词吧！

第112-114页
动物词语的词源

这部分介绍了许多动物词语的来源，如 aardvark（土豚）和 wildebeest（角马），还介绍了这些单词在日常话语中的用法。

第115页
动物习语

习语是指有固定搭配的短语或词组，意思与字面不同。本书的这部分介绍了与动物有关的常用习语。

第116-117页
动物知识小测验

这部分是趣味小测验，可以用来测试自己学会了多少动物知识。小测验的所有答案都能在书中找到，所以请仔细看书，看看自己能否全都答对！小测验的答案在第131—132页。

Animal life 动物总览

There are millions of species of animals on Earth and they come in an amazing range of shapes and sizes! Some tiny creatures can only be seen with the help of a microscope. The largest animal on Earth is the blue whale.

地球上有几百万种动物，它们的形态和体形千差万别。有的体形微小，只有在显微镜下才能看见。世界上最大的动物是蓝鲸。

Vertebrates and invertebrates 脊椎动物和无脊椎动物

Animals can be vertebrates or invertebrates. Vertebrates have a spine (or backbone). Invertebrates do not have a spine. Some invertebrates have soft bodies, and some have a hard outer covering.

动物分脊椎动物和无脊椎动物。脊椎动物有脊椎（又称脊柱），无脊椎动物没有脊椎。有的无脊椎动物身体柔软，有的有坚硬的外壳。

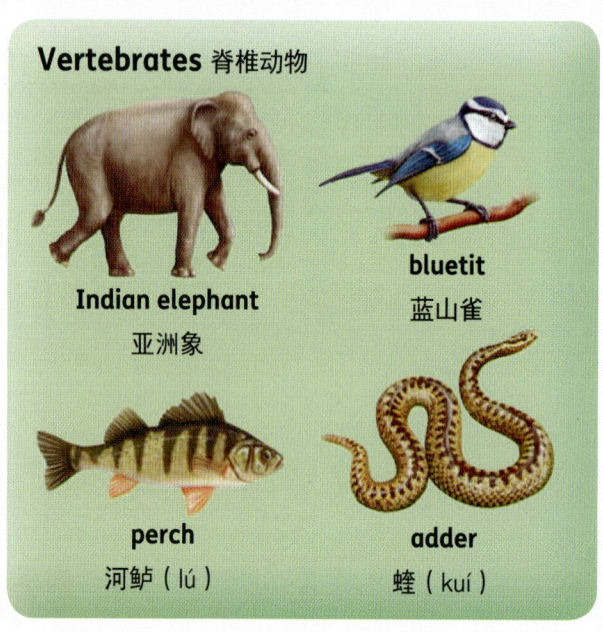

Vertebrates 脊椎动物
- Indian elephant 亚洲象
- bluetit 蓝山雀
- perch 河鲈（lú）
- adder 蝰（kuí）

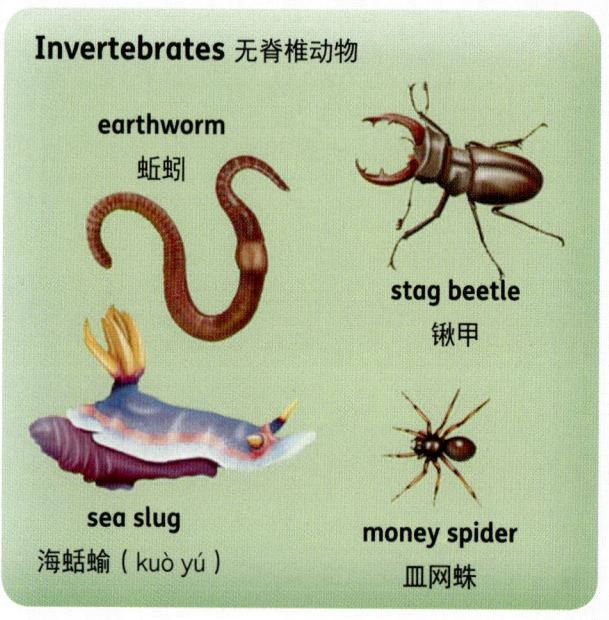

Invertebrates 无脊椎动物
- earthworm 蚯蚓
- stag beetle 锹甲
- sea slug 海蛞蝓（kuò yú）
- money spider 皿网蛛

Moving and eating 移动与进食

Animals are different from other living things, such as plants and fungi, because they are more mobile and they survive by eating other life forms. They can be divided into three groups, according to what they eat. Carnivores eat meat. Herbivores eat plants. Omnivores eat meat and plants.

与植物、真菌等其他生物不同，动物可以更方便地活动，靠吃其他生物存活。根据食物不同，动物可分三类：食肉动物吃肉，食草动物吃植物，杂食动物既吃肉也吃植物。

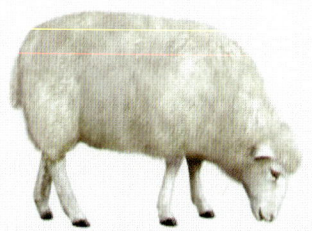

Sheep graze on grass. They are herbivores.
绵羊吃草，是食草动物。

Tigers hunt and kill their prey. They are carnivores.
老虎捕杀猎物，是食肉动物。

8

Animal life 动物总览

Parasites 寄生物

Parasites feed off the body of another animal. The animal that a parasite feeds off is known as the "host".
寄生物从其他动物身上获取养分，寄生物所寄生的动物叫"宿主"。

Tapeworms live inside the guts of other animals.
绦（tāo）虫寄生在其他动物的肠道里。

Leeches suck blood from larger creatures.
水蛭（zhì）吸食体形更大的动物的血液。

Fleas live in the coats of mammals and birds.
跳蚤住在哺乳动物的皮毛和鸟类的羽毛中。

Animal senses 动物感官

Animals use their senses of sight, hearing, smell and touch to help them to find their food and to stay safe. Different animals rely on different senses.
动物利用视觉、听觉、嗅觉和触觉寻找食物，躲避危险。不同的动物依靠不同的感官。

Owls have large eyes that let in a lot of light so they can hunt at night.
猫头鹰有一双大眼睛，能吸纳更多光线，所以能在晚上捕猎。

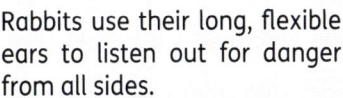

Mice use their whiskers to check if they can fit through a space.
老鼠用胡须来判断自己能否钻过缝隙。

A wolf's powerful sense of smell allows it to track an animal's scent.
狼靠敏锐的嗅觉来追踪猎物。

Rabbits use their long, flexible ears to listen out for danger from all sides.
兔子用灵活的长耳朵监听四周的声音，预防危险。

9

本章音频

Animal behaviour 动物行为

Scientists who study animals are called zoologists. They observe animals very closely and study their behaviour. These two pages cover a range of animal behaviour: courtship, fighting, migration and hibernation.

研究动物的科学家叫动物学家,他们仔细观察动物,研究它们的行为。这两页描述的动物行为有求偶、打斗、迁徙和冬眠。

Courtship 求偶

Most creatures need to find a mate in order to create new life. Males use a wide range of courtship behaviour to help them win a mate. Courtship rituals are designed to attract the attention of a female and to show that a male is better than his rivals.

大多数动物都需要寻找配偶以繁衍后代。雄性动物的求偶方式多种多样。求偶仪式的目的是为了吸引雌性动物,显示自己强于竞争对手。

The bowerbird collects colourful objects.
园丁鸟收集颜色鲜艳的物件。

The fiddler crab waves its enormous claw.
招潮蟹挥动巨大的螯(áo)。

Tree frogs make repeated loud croaks.
树蛙反复大声鸣叫。

The bird of paradise displays its plumage.
极乐鸟展示羽毛。

Fighting 打斗

Males often have to fight off other rivals before they can win their mate. This behaviour means the strongest males win the best mates.

雄性动物常常需要打败对手才能赢得配偶,这一行为意味着最强壮的雄性动物能赢得最佳的配偶。

Stags lock antlers to fight.
雄鹿抵角相斗。

Rattlesnakes try to push each other to the ground.
响尾蛇把对手压向地面。

10

Animal behaviour 动物行为

Migration 迁徙

Some animals make very long journeys to breed or find food. This is called migration. Many migrating creatures make the same journey each year. They always travel around the same time of the year.

有些动物为繁殖或觅食而长途旅行，这种行为叫作迁徙。许多迁徙动物每年的旅行路线都相同，并且每年差不多都在同一时间开始旅行。

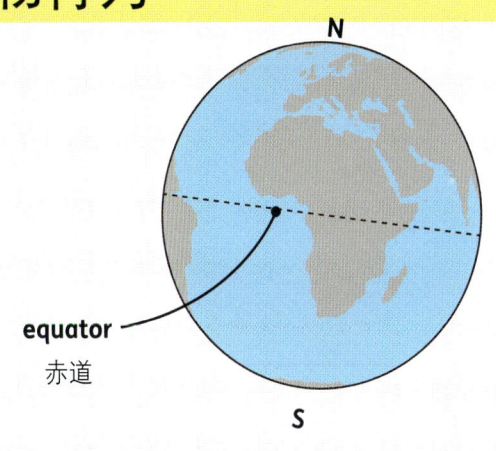

equator
赤道

Barn swallows avoid the cold northern winter by migrating south to warmer regions.

家燕飞向南方的温暖地带，躲避北方寒冷的冬天。

Atlantic salmon are born in rivers, migrate to the ocean, and then return to the same river to breed.

大西洋鲑（guī）出生在河流中，它们会游向大海，再返回同一条河里产卵。

Humpback whales give birth near the equator, but their main feeding grounds are in the polar seas.

座头鲸在赤道附近产崽，而它们的主要觅食区在极地海域。

Hibernation 冬眠

In the cold winter months some animals hibernate. They sink into a very deep sleep and all their body functions slow down. Before they start to hibernate, some animals consume a large amount of food. This helps to keep them strong while they are hibernating.

在寒冷的冬季，有的动物会冬眠。它们进入深层的休眠状态，所有的身体功能都放缓。在开始冬眠之前，有的动物会大量进食。这能帮助它们在冬眠期间保持健康。

pipistrelle bat
伏翼，油蝠

box turtle
箱龟

hamster
仓鼠

bumblebee
熊蜂

11

An amazing range of creatures lives on Earth. This section introduces the main animal groups: mammals; birds; reptiles; amphibians; fish and other sea creatures; and insects and minibeasts.

地球上的动物种类数目惊人。本部分介绍主要的动物类群：哺乳动物、鸟类、爬行动物、两栖动物、鱼类和其他海洋动物、昆虫和其他无脊椎动物。

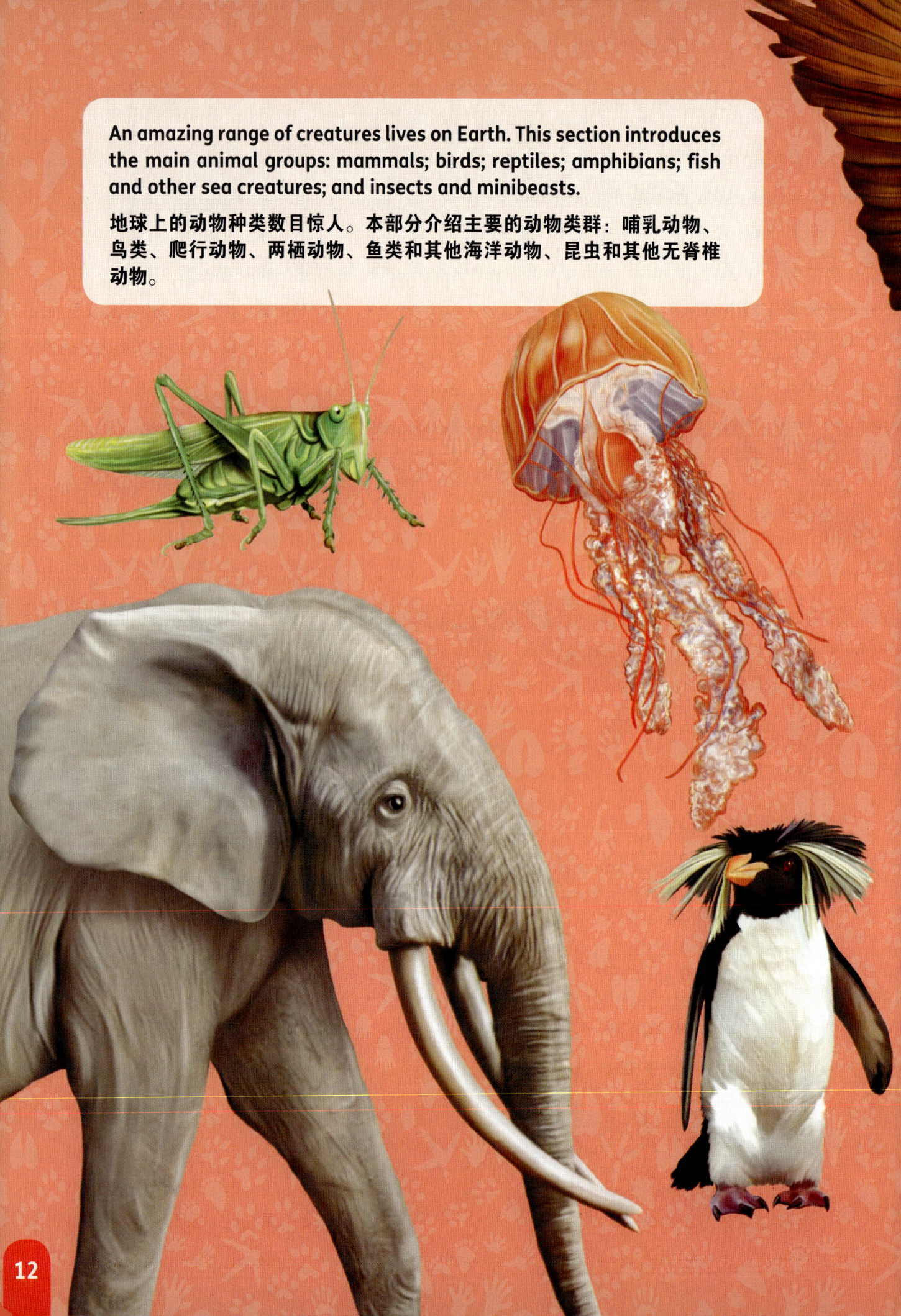

ALL KINDS OF ANIMALS
各类动物

All kinds of animals 各类动物

Mammals 哺乳动物

Mammals range in size from tiny dormice to enormous elephants. They include bats and whales and, of course, human beings! All mammals are warm-blooded (which means their bodies always stay warm). They all have fur, bristles or body hair and all mammal mothers produce milk to feed their young.

从小小的睡鼠到庞大的大象，哺乳动物大小各异。蝙蝠、鲸，还有人类，都是哺乳动物！所有哺乳动物都是温血动物（它们的身体能保持稳定的体温）。它们有皮毛、刚毛或毛发，所有哺乳动物母亲都会分泌乳汁喂养幼崽。

Horse 马
- mane 鬃毛
- nostril 鼻孔
- foreleg 前腿
- hoof 蹄
- fetlock 球节
- barrel 躯体
- flank 胁腹
- hind leg 后腿
- croup 臀部

Dog 狗
- ear 耳朵
- tail 尾巴
- back 背部
- paw 脚掌
- tongue 舌头
- teeth 牙齿
- muzzle 口鼻
- eye 眼睛
- claw 爪

All kinds of animals 各类动物

Mammals 哺乳动物

Monkey 猴

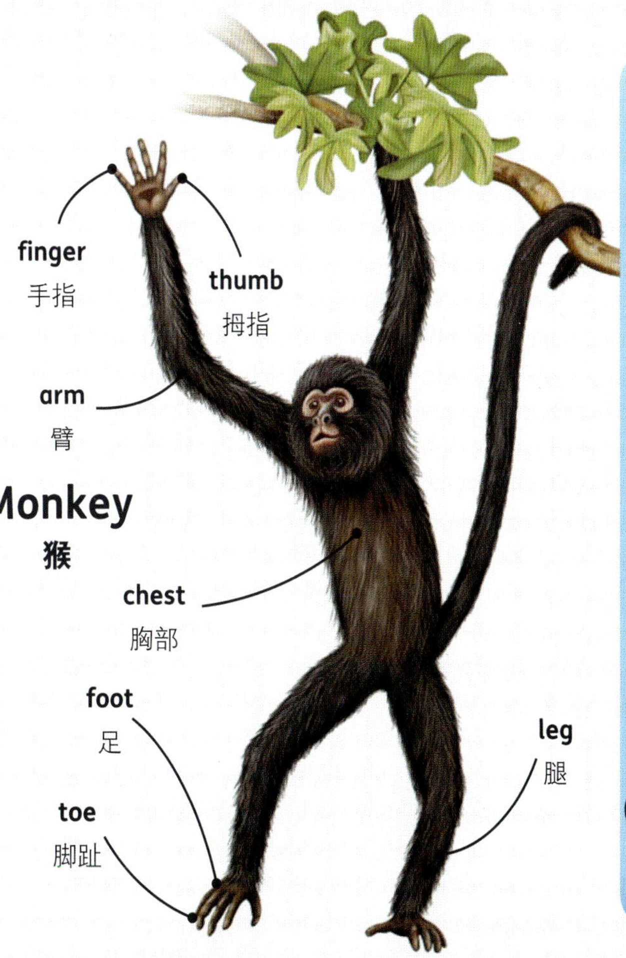

- finger 手指
- thumb 拇指
- arm 臂
- chest 胸部
- foot 足
- toe 脚趾
- leg 腿

Internal organs of an ape
类人猿的内脏

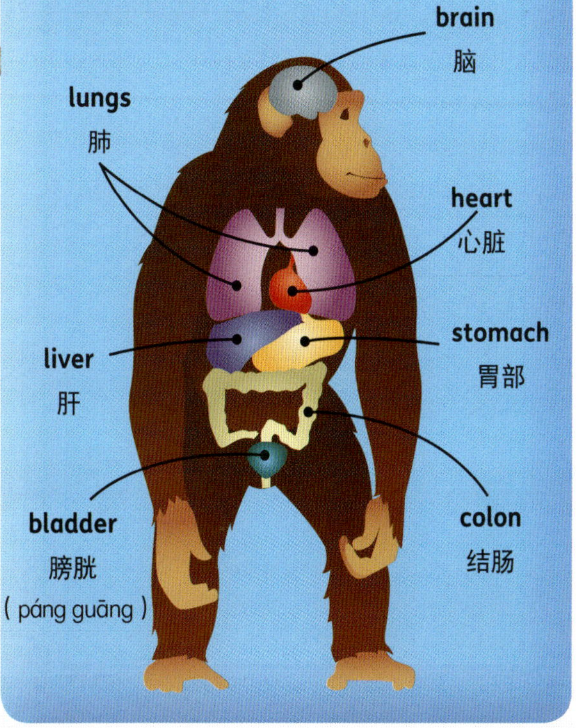

- brain 脑
- lungs 肺
- heart 心脏
- liver 肝
- stomach 胃部
- bladder 膀胱 (páng guāng)
- colon 结肠

Marsupials 有袋类动物

Marsupials belong to a small group of mammals that are mainly found in Australia. The mothers carry their babies in a pouch until the babies are old enough to survive on their own.

有袋类动物是哺乳动物中的一小类，主要生活在澳大利亚。雌性有袋类动物将幼崽装在育儿袋中，直到它们能够独立存活。

Marsupials living in Australia include kangaroos, wallabies and wombats. A few species of marsupials are found in North and South America. They include opossums and shrew opossums.

澳大利亚的有袋类动物有袋鼠、沙袋鼠、毛鼻袋熊等等。在南北美洲也有一些有袋类动物，如负鼠和鼩 (qú) 负鼠。

Kangaroo 袋鼠

- baby 幼崽
- teat 乳头
- pouch 育儿袋

The baby suckles milk from its mother's teats.
袋鼠幼崽从母亲的乳头吸取乳汁。

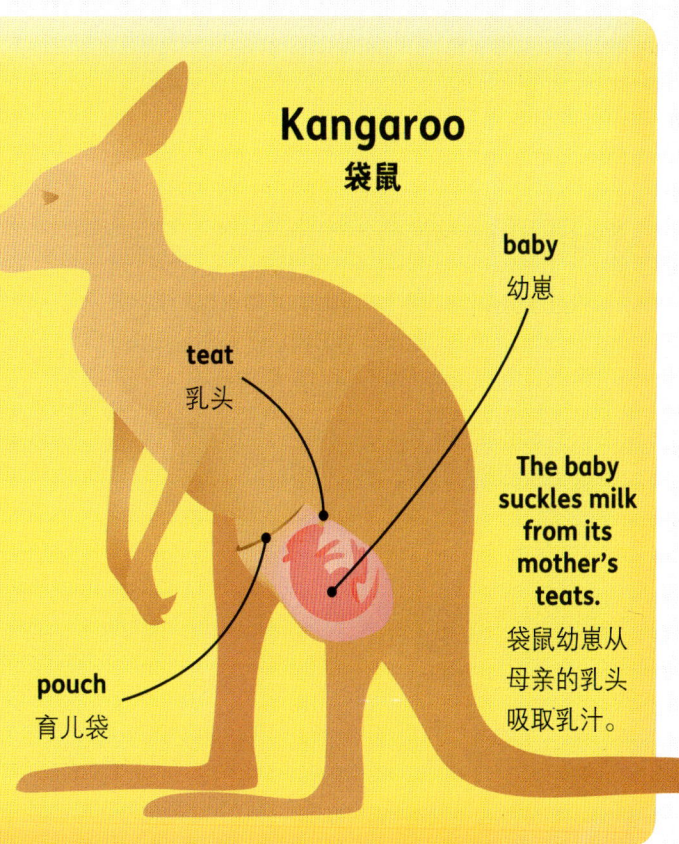

All kinds of animals 各类动物

Birds 鸟类

Birds are warm-blooded and covered with feathers. All birds lay eggs and most of them can fly, although there are a few large flightless birds, such as the ostrich and the penguin. Many birds perch in trees and spend most of their time in the air. Some birds can swim and live mainly on water.

鸟类是温血动物，体表覆盖着羽毛。所有鸟类都产卵，大部分鸟会飞，也有少数不会飞的大鸟，如鸵鸟和企鹅。许多鸟类在树上栖息，大部分时间在空中飞翔。有的鸟会游泳，主要在水上生活。

wingtip 翅尖

beak 喙（huì）

Eagle 雕

wing 翅膀

talons 爪

tail 尾羽

Feathers 羽毛

Many birds have three types of feather. Each type of feather has a different shape and each has its own important function.

许多鸟类有三种羽毛，每种羽毛形状不同，各有各的重要功能。

Body feathers give the bird a streamlined shape.
正羽使鸟类的身躯呈流线型。

Down feathers keep the bird's body warm.
绒羽使鸟类的身体保持温暖。

Wing and tail feathers allow it to fly.
飞羽和尾羽使鸟类得以飞翔。

All kinds of animals 各类动物

Different bills 不同的鸟嘴

Bills are sometimes called beaks. They have a range of different shapes that are suited to different ways of feeding.

鸟嘴又称鸟喙，在英语中叫作 bill 或 beak。鸟喙的形状各异，适应不同的进食方式。

Waders grab fish underwater.
涉禽从水中抓鱼。

Birds of prey tear up their food.
猛禽将猎物撕碎。

Parrots crack and shell nuts.
鹦鹉咬碎坚果壳。

Ducks scoop up food from the water's surface.
鸭子用嘴在水面上舀取食物。

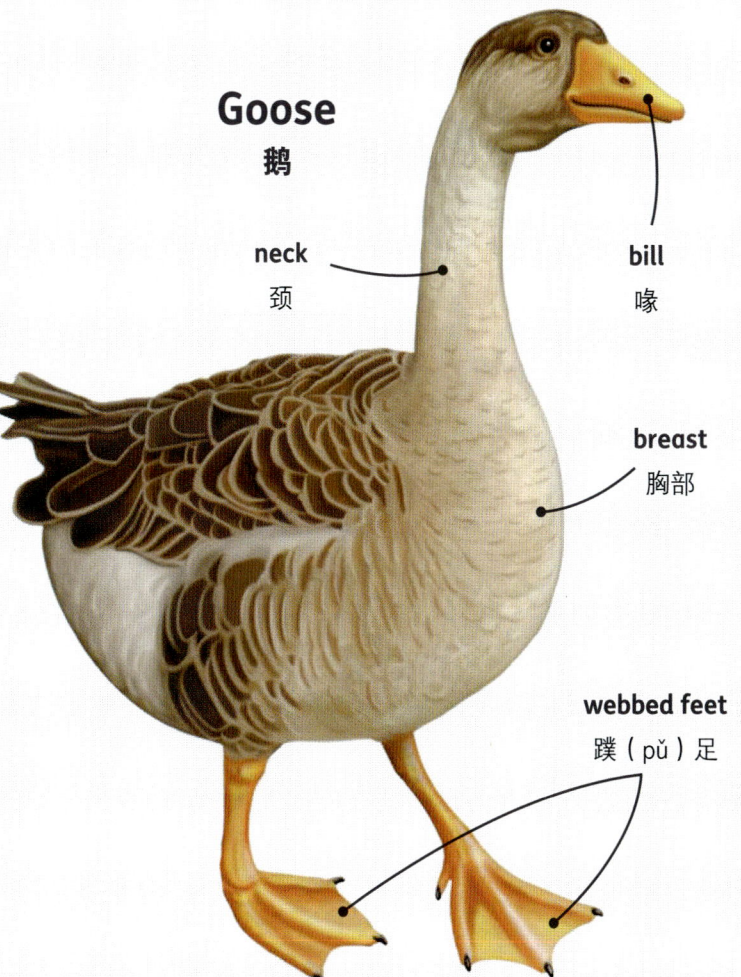

Goose 鹅

- neck 颈
- bill 喙
- breast 胸部
- webbed feet 蹼（pǔ）足

Birds 鸟类

17

All kinds of animals 各类动物

Reptiles 爬行动物

Reptiles have a dry, tough skin that is covered with scales. All reptiles are cold-blooded, so they need to bask in the sun to warm up. Reptiles include crocodiles, lizards, turtles and snakes.

爬行动物的皮肤干燥粗糙，布满鳞片。所有爬行动物都是冷血动物，它们需要晒太阳来保持温暖。鳄鱼、蜥蜴（xī yì）、龟和蛇都是爬行动物。

Crocodile 鳄鱼
- nostril 鼻孔
- snout 吻
- scaly skin 鳞状皮肤
- jaw 颌（hé）
- claws 爪

Turtle 海龟
- flipper 鳍足
- shell 壳
- underbelly 下腹部

All kinds of animals 各类动物

Reptiles 爬行动物

Camouflage 保护色

Many animals use camouflage to help them blend in with their surroundings. Their colouring makes them hard for hunters to spot.

许多动物利用保护色隐藏在环境中，它们的颜色使天敌很难发现它们。

Burmese python 缅甸蟒（mǎng）

Chinese water dragon lizard 长鬣（liè）蜥

Hermann's tortoise 赫氏陆龟

Snake 蛇

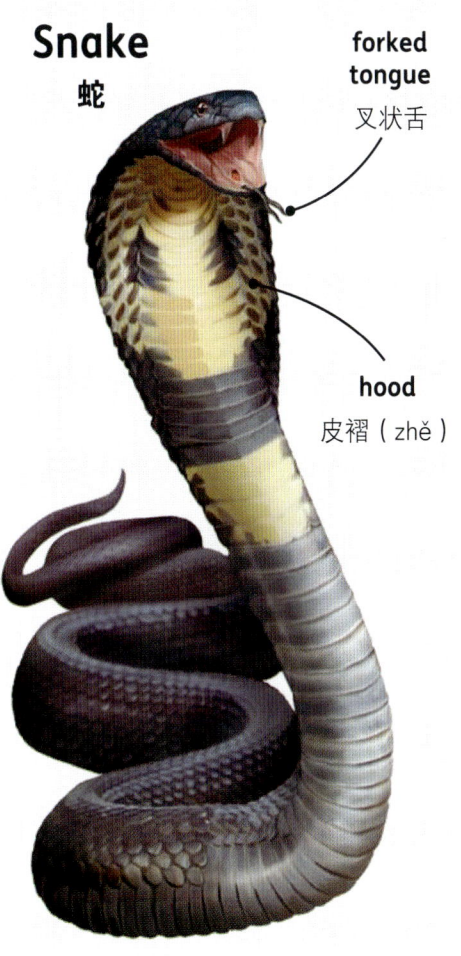

- forked tongue 叉状舌
- hood 皮褶（zhě）

Venom 毒液

Some snakes and lizards inject a poison called venom into their prey. They use their fangs to puncture a victim's skin and then inject their poison. Venom can paralyse or even kill a victim.

有些蛇和蜥蜴将毒液注入猎物体内。它们用毒牙咬破猎物的皮肤，注入毒液。毒液能麻痹甚至杀死猎物。

Viper 蝰蛇

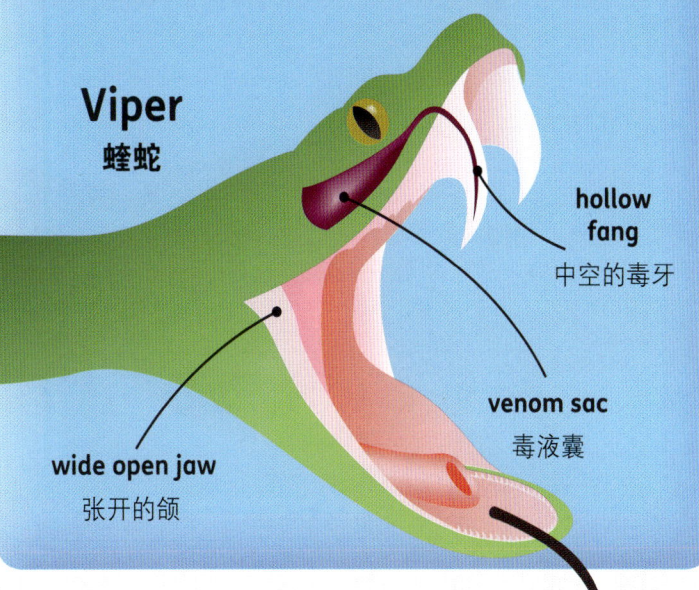

- hollow fang 中空的毒牙
- venom sac 毒液囊
- wide open jaw 张开的颌

19

All kinds of animals 各类动物

Amphibians 两栖动物

Amphibians have smooth skin, with no scales or hair. They are cold-blooded. They usually hatch and develop in water, but then spend most of their adult lives on land. Frogs, toads, salamanders and newts are all amphibians. There are also some amphibians that look like large worms.
两栖动物表皮光滑，没有鳞片或毛发。它们是冷血动物，一般在水中出生和发育，成年后大部分时间生活在陆地上。蛙、蟾蜍（chán chú）和蝾螈（róng yuán）都是两栖动物。还有一些两栖动物长得像大蠕虫。

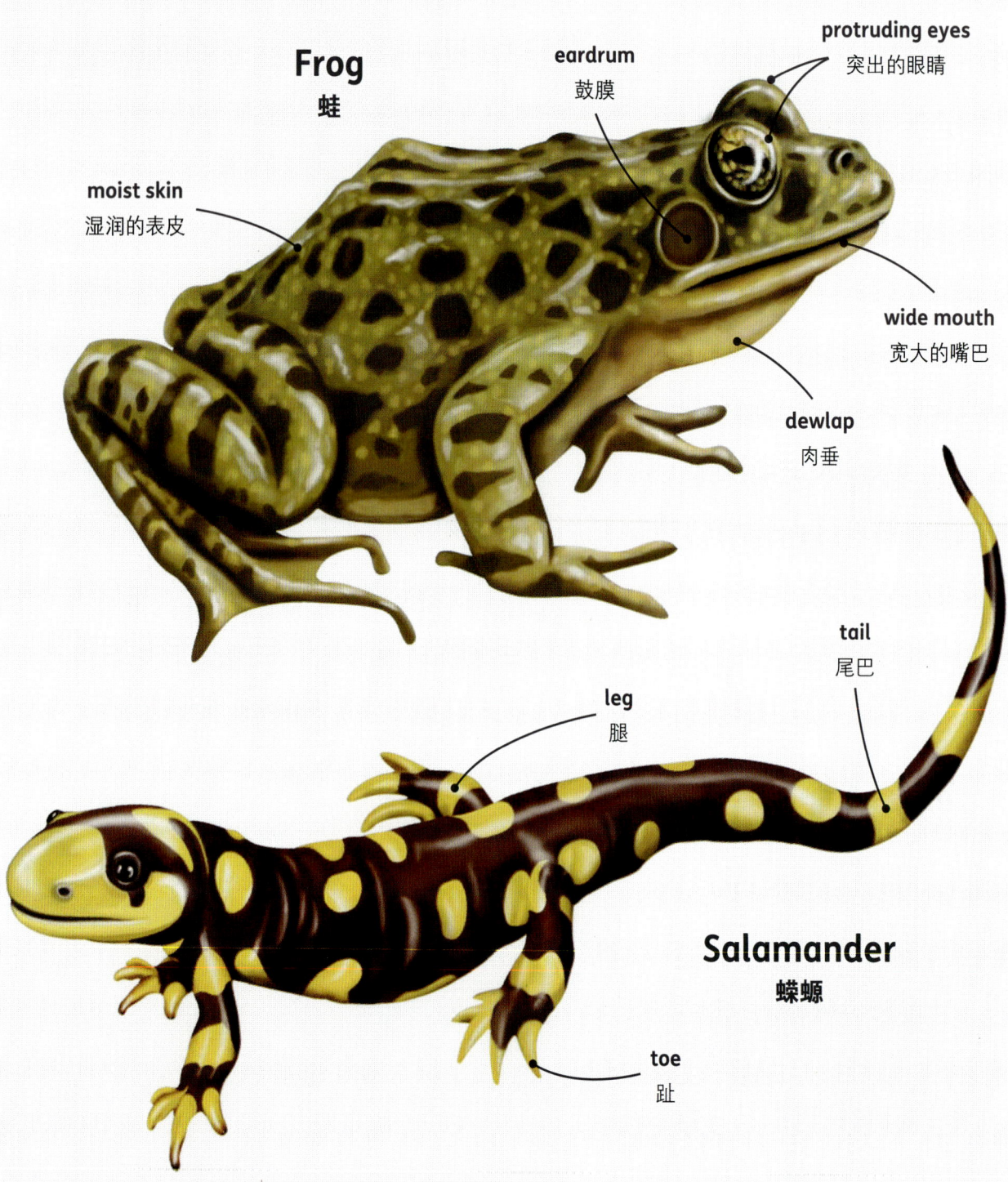

Frog 蛙
- eardrum 鼓膜
- protruding eyes 突出的眼睛
- moist skin 湿润的表皮
- wide mouth 宽大的嘴巴
- dewlap 肉垂

Salamander 蝾螈
- leg 腿
- tail 尾巴
- toe 趾

All kinds of animals 各类动物

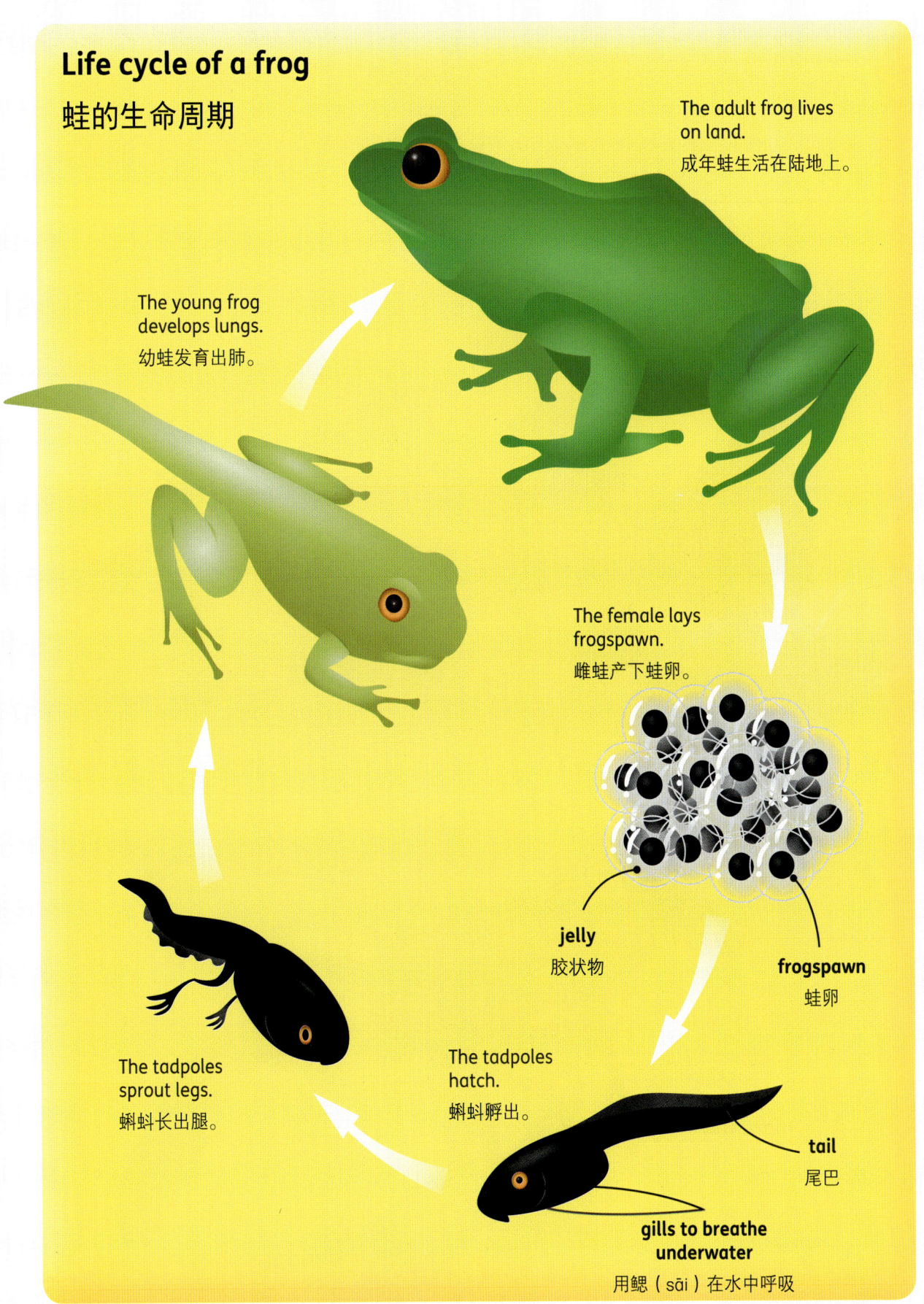

Life cycle of a frog
蛙的生命周期

The adult frog lives on land.
成年蛙生活在陆地上。

The young frog develops lungs.
幼蛙发育出肺。

The female lays frogspawn.
雌蛙产下蛙卵。

jelly
胶状物

frogspawn
蛙卵

The tadpoles sprout legs.
蝌蚪长出腿。

The tadpoles hatch.
蝌蚪孵出。

tail
尾巴

gills to breathe underwater
用鳃（sāi）在水中呼吸

Amphibians 两栖动物

All kinds of animals 各类动物

Fish and other sea creatures 鱼类和其他海洋动物

Fish are creatures with backbones that are specially adapted to live underwater. Other sea creatures include sponges, worms, jellyfish, starfish, squid and lobsters. Many sea creatures are microscopic—they can only be seen with the help of a microscope.

鱼类有脊柱，适应水中生活。其他海洋动物有海绵、海洋蠕虫、水母、海星、乌贼和龙虾等等。许多海洋动物非常微小——只有借助显微镜才能看到。

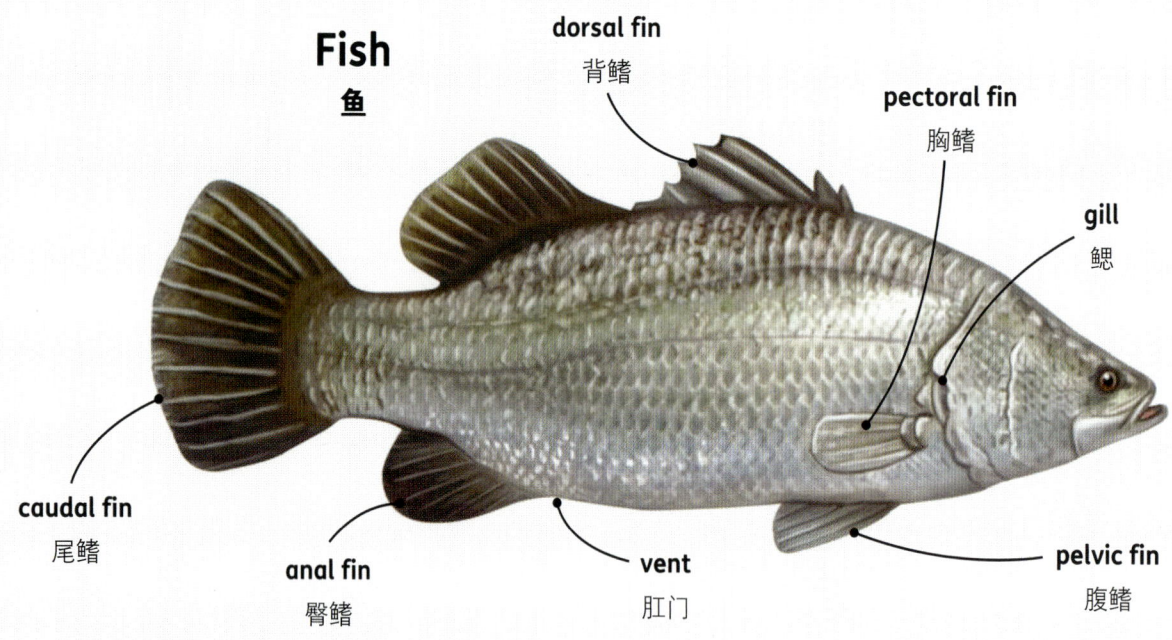

Fish 鱼
- dorsal fin 背鳍
- pectoral fin 胸鳍
- gill 鳃
- pelvic fin 腹鳍
- vent 肛门
- anal fin 臀鳍
- caudal fin 尾鳍

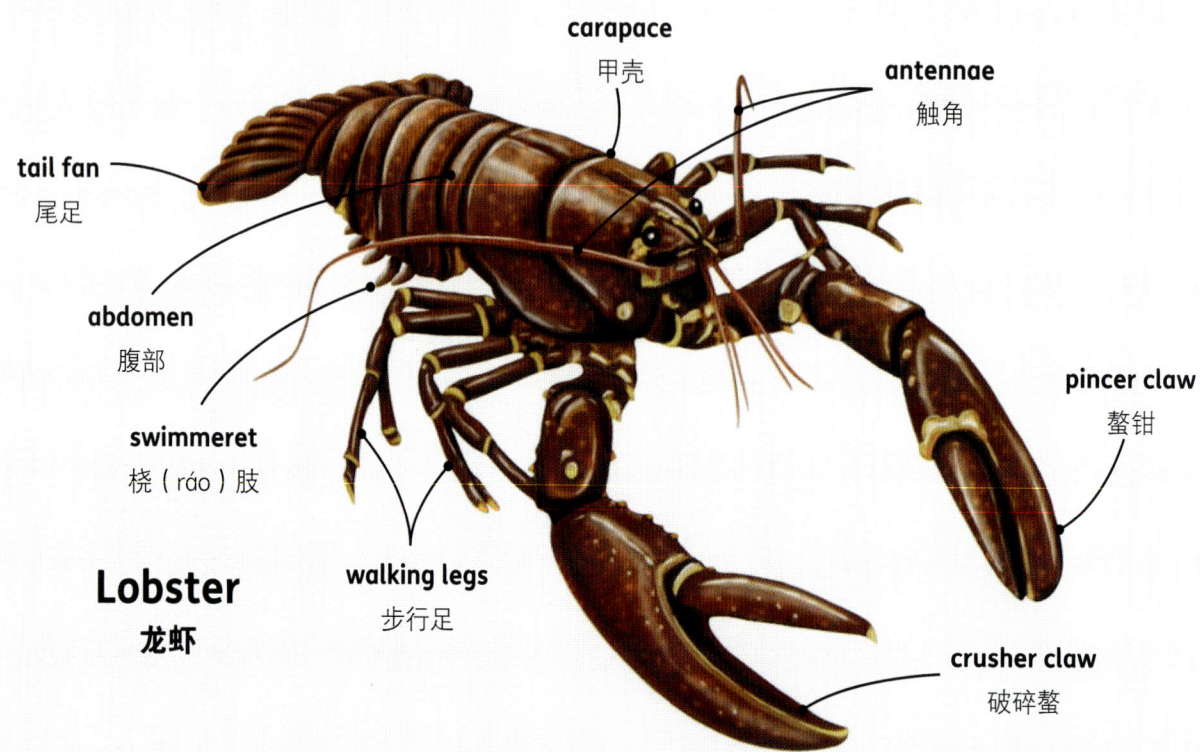

Lobster 龙虾
- carapace 甲壳
- antennae 触角
- pincer claw 螯钳
- crusher claw 破碎螯
- walking legs 步行足
- swimmeret 桡（ráo）肢
- abdomen 腹部
- tail fan 尾足

All kinds of animals 各类动物

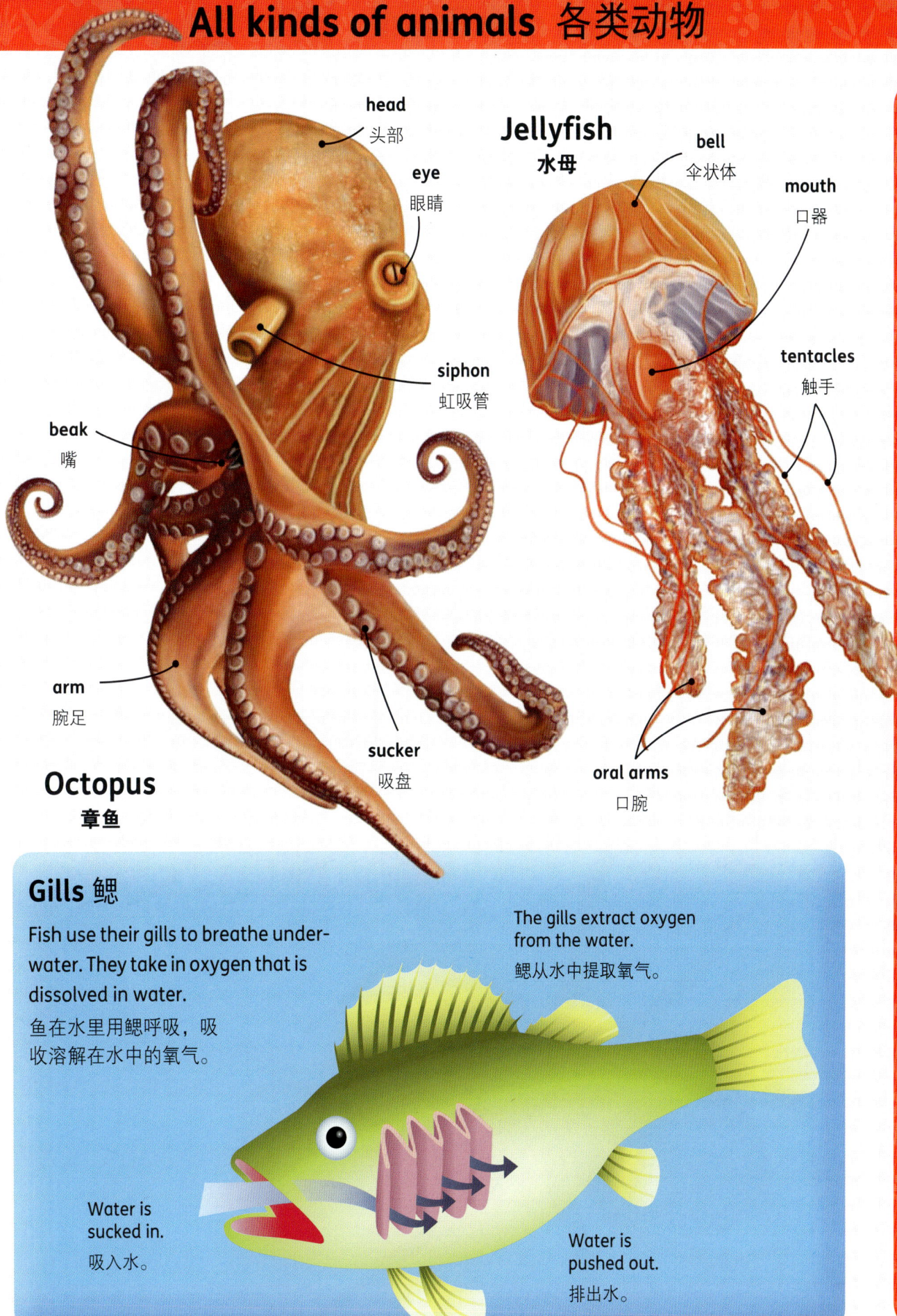

Jellyfish 水母
- bell 伞状体
- mouth 口器
- tentacles 触手
- oral arms 口腕

Octopus 章鱼
- head 头部
- eye 眼睛
- siphon 虹吸管
- beak 嘴
- arm 腕足
- sucker 吸盘

Gills 鳃

Fish use their gills to breathe underwater. They take in oxygen that is dissolved in water.
鱼在水里用鳃呼吸，吸收溶解在水中的氧气。

The gills extract oxygen from the water.
鳃从水中提取氧气。

Water is sucked in.
吸入水。

Water is pushed out.
排出水。

Fish and other sea creatures 鱼类和其他海洋动物

23

All kinds of animals 各类动物

Insects and minibeasts 昆虫和其他无脊椎动物

Insects have bodies with three parts: the head, the thorax and the abdomen. All insects have six legs, and some have wings. Insects include beetles, butterflies and bees. Minibeasts are small creatures without a backbone. They include spiders, centipedes and woodlice.

昆虫的身体由三部分组成：头部、胸部、腹部。所有昆虫都有六条腿，有的昆虫有翅膀。甲虫、蝴蝶和蜜蜂等都是昆虫。其他小型无脊椎动物是没有脊柱的小动物，如蜘蛛、蜈蚣、鼠妇等等。

Insects make up the biggest group of living creatures. There are over 900 thousand different species of insects, and eight out of ten of all the world's creatures are insects.

昆虫是所有动物中数量最多的类群。昆虫的种类超过 90 万种，数量占全世界动物总数的八成。

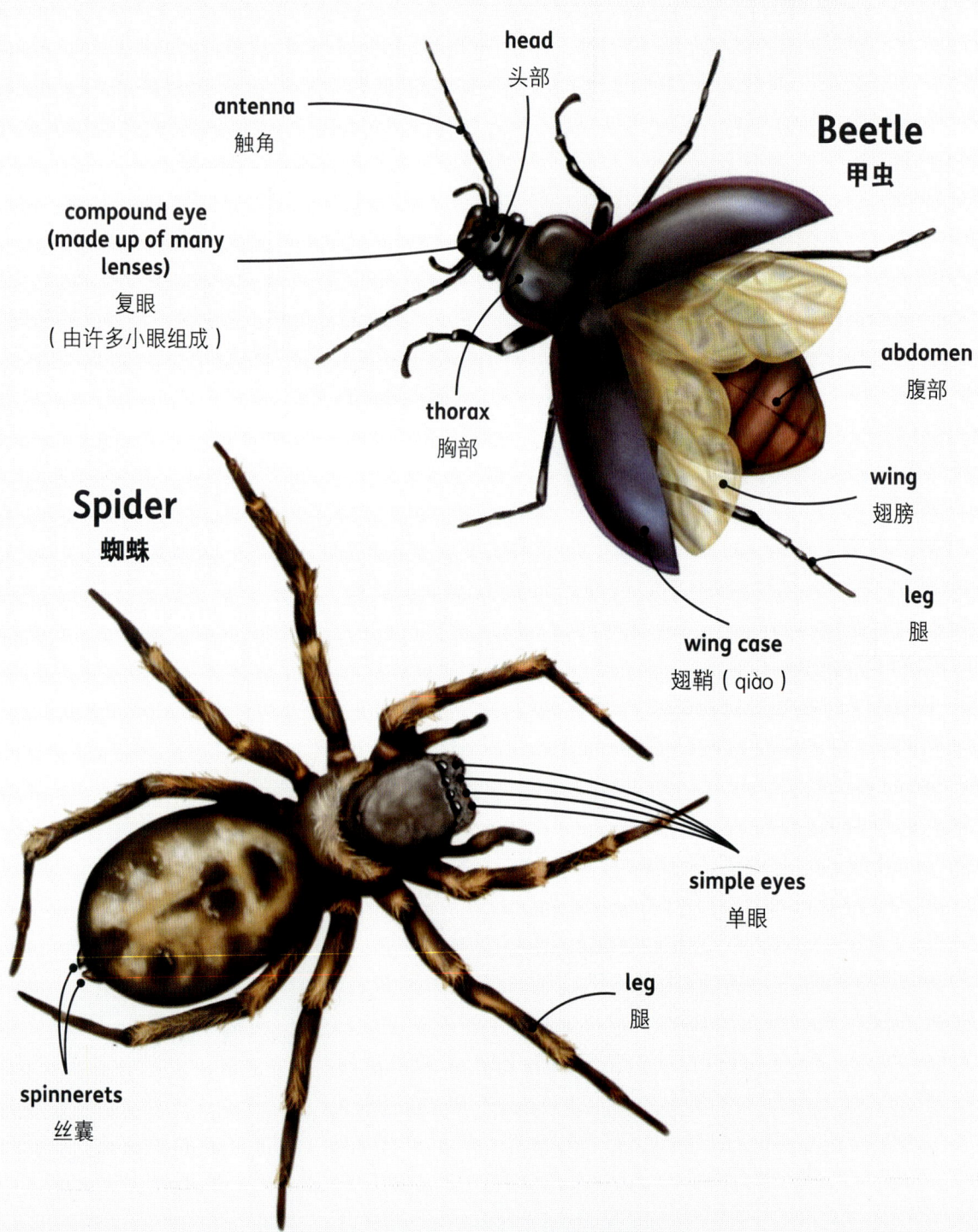

All kinds of animals 各类动物

Honey bee 蜜蜂

The honey bee collects nectar from inside flowers and spreads pollen from one flower to another.

蜜蜂从花朵中采集花蜜，并在花朵之间传播花粉。

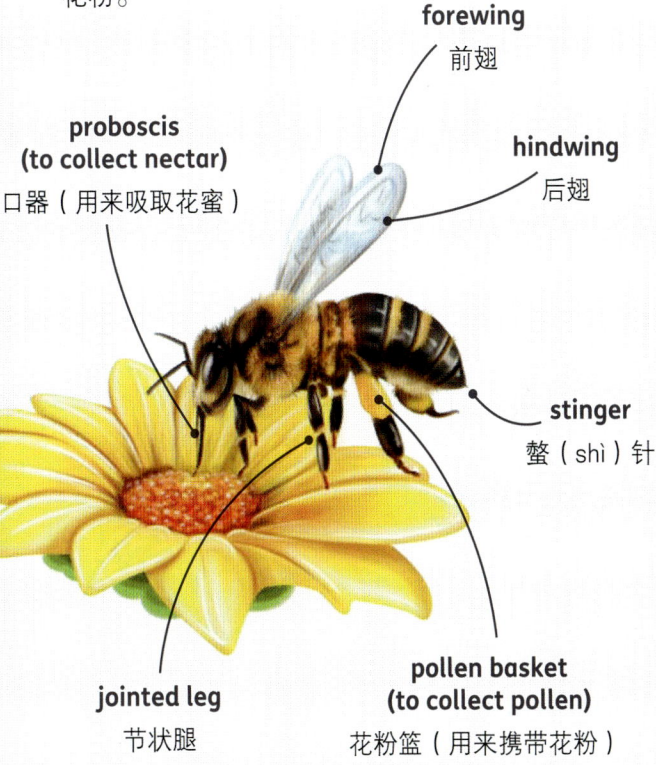

- **forewing** 前翅
- **hindwing** 后翅
- **proboscis (to collect nectar)** 口器（用来吸取花蜜）
- **stinger** 螫（shì）针
- **jointed leg** 节状腿
- **pollen basket (to collect pollen)** 花粉篮（用来携带花粉）

Beehives 蜂窝

A beehive is home to a colony of bees. The colony is made up of a single queen bee, hundreds of drones and thousands of workers.

蜂窝是蜂群居住的地方。蜂群由蜂王、几百只雄蜂和几千只工蜂组成。

queen 蜂王
The queen lays eggs to produce more bees.
蜂王产卵，孵出更多蜜蜂。

drone 雄蜂
Drones mate with the queen to start new hives.
雄蜂与蜂王交配，繁衍新的蜂群。

worker 工蜂
Workers look after the young and make honey.
工蜂照顾幼虫，酿造蜂蜜。

Life cycle of a butterfly
蝴蝶的生命周期

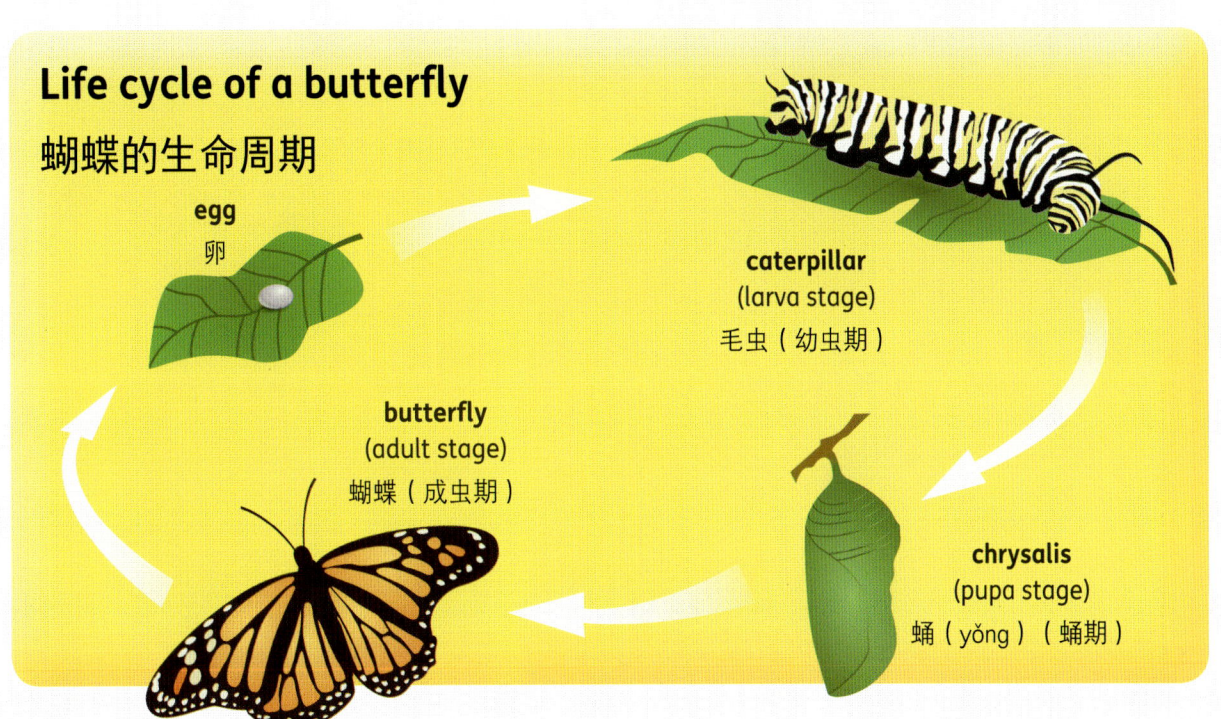

- **egg** 卵
- **caterpillar (larva stage)** 毛虫（幼虫期）
- **chrysalis (pupa stage)** 蛹（yǒng）（蛹期）
- **butterfly (adult stage)** 蝴蝶（成虫期）

Insects and minibeasts 昆虫和其他无脊椎动物

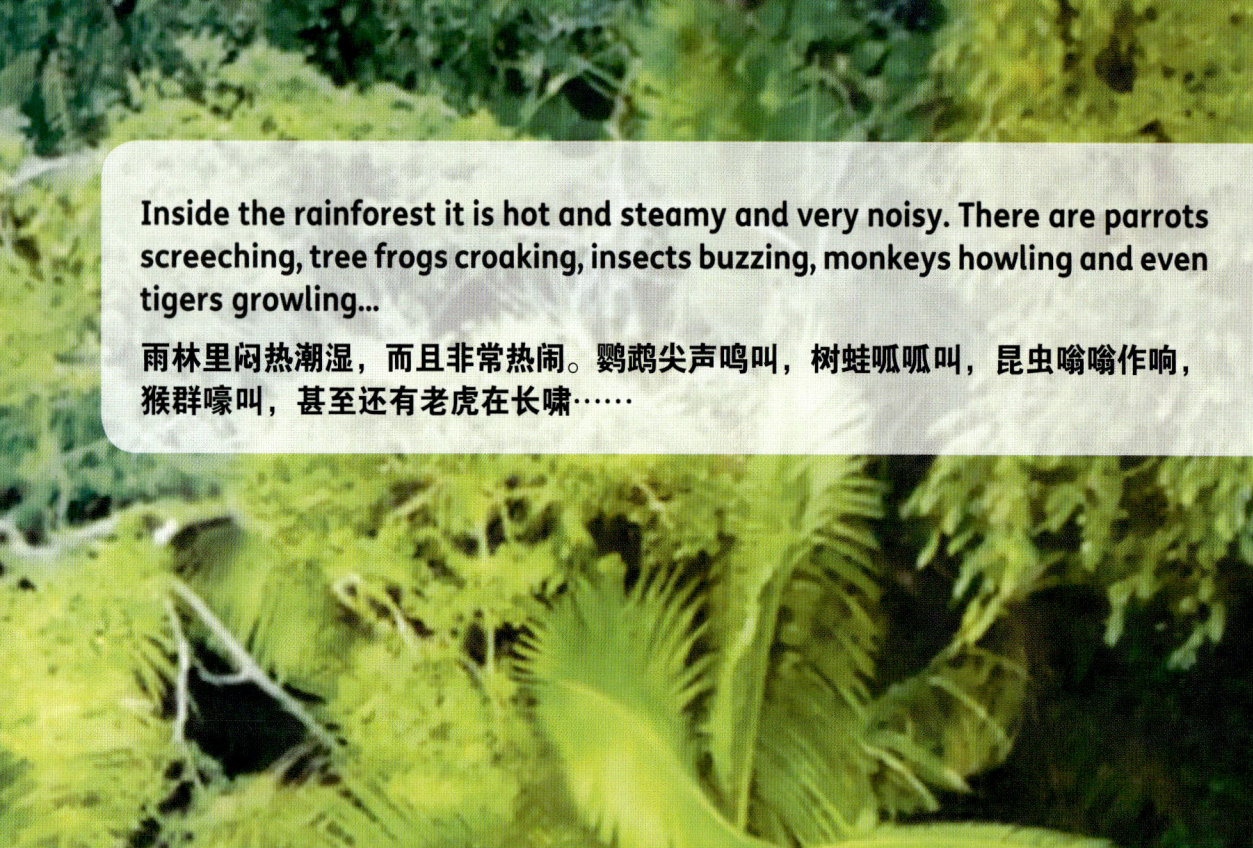

Inside the rainforest it is hot and steamy and very noisy. There are parrots screeching, tree frogs croaking, insects buzzing, monkeys howling and even tigers growling...

雨林里闷热潮湿,而且非常热闹。鹦鹉尖声鸣叫,树蛙呱呱叫,昆虫嗡嗡作响,猴群嚎叫,甚至还有老虎在长啸……

RAINFOREST CREATURES
雨林动物

本章音频

Rainforest creatures 雨林动物

Rainforest habitats 雨林栖息地

Most rainforests are tropical. They grow close to the equator, where it is hot and rainy all year round. Temperate rainforests grow in cooler parts of the world. They are further from the equator, and are often found along the coast. Rainforests are home to an astonishing range of species. The world's largest rainforest is the Amazon in South America. One in ten of all the world's known species live in the Amazon rainforest.

多数雨林位于热带地区，也就是赤道附近，那里常年湿热多雨。温带雨林的生长环境则要凉爽一些，离赤道较远，多沿海岸分布。雨林孕育了各色各样的物种。世界上最大的雨林位于南美洲的亚马孙，十分之一世界已知物种生活在这片雨林之中。

Rainforest layers 雨林层级

Rainforests have four main layers. Each one is home to different animals, although some creatures move between the layers.

雨林通常可分为四层，每一层都有不同的动物，但有些动物也会在不同层级之间活动。

Emergent layer 林上层 — morpho butterfly 闪蝶

Canopy 树冠层 — flying dragon lizard 飞蜥 — great hornbill 双角犀鸟

Understorey 林下层 — red-eyed tree frog 红眼树蛙 — clouded leopard 云豹

Forest floor 林地层 — tarantula 狼蛛 — giant anteater 大食蚁兽

Rainforest creatures 雨林动物

Major rainforest regions 主要雨林分布图

- tropical rainforest 热带雨林
- temperate rainforest 温带雨林

harpy eagle 美洲角雕

sugar glider 蜜袋鼯（wú）

sunbird 太阳鸟

chimpanzee 黑猩猩

gibbon 长臂猿

emerald tree boa 绿树蚺（rán）

goliath beetle 巨花潜金龟

coral snake 珊瑚蛇

Rainforest habitats 雨林栖息地

Rainforest creatures 雨林动物

Life on the rainforest floor 林地层动物

Very little light reaches the forest floor. Insects and reptiles live among leaves and vines, while mammals and flightless birds weave their way through the tree trunks. Some of the creatures that prowl the forest floor are very large. Elephants, gorillas and tigers all live in the rainforest.

光线很难照射到林地层。昆虫和爬行动物在树叶和藤蔓间生活，哺乳动物和不会飞的鸟类在林间穿行。有的林地层动物很大。大象、大猩猩、老虎都生活在雨林中。

There are also some giant insects, such as the Hercules beetle, which measures around 15cm (6 inches) and can lift eight times its own weight!

还有一些巨大的昆虫，比如巨犀金龟，身长约 15 厘米（6 英寸），能举起重量是自身重量 8 倍的物体！

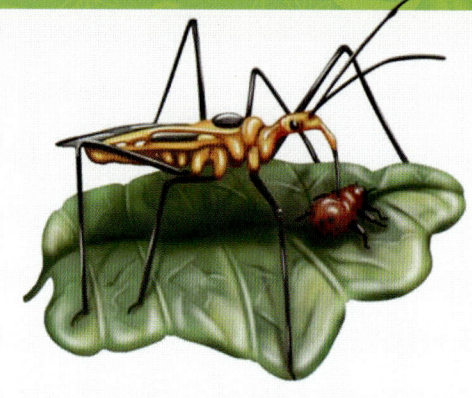

assassin bug 猎蝽（chūn）
Assassin bugs inject their prey with a deadly saliva that turns body contents to liquid.
猎蝽将致命的唾液注入猎物身体，把猎物的身体化为液体。

tapir 貘（mò）

jaguar 美洲豹
Jaguars prowl through the undergrowth and also climb up into the understorey.
美洲豹在灌木丛中潜行，有时也爬到林下层。

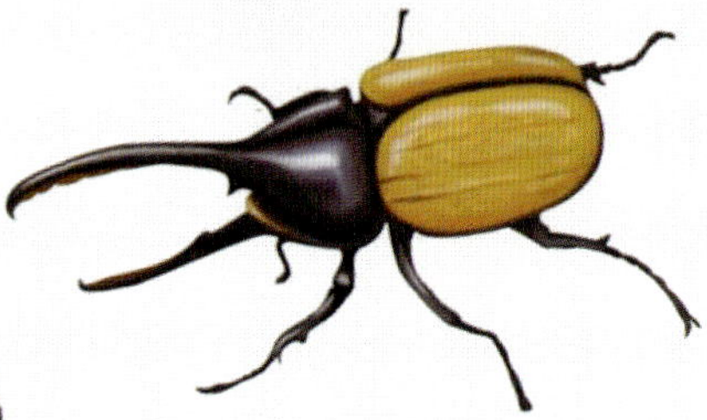

Hercules beetle
巨犀金龟

funnel web spider
漏斗网蛛

jewel beetle
彩虹吉丁虫

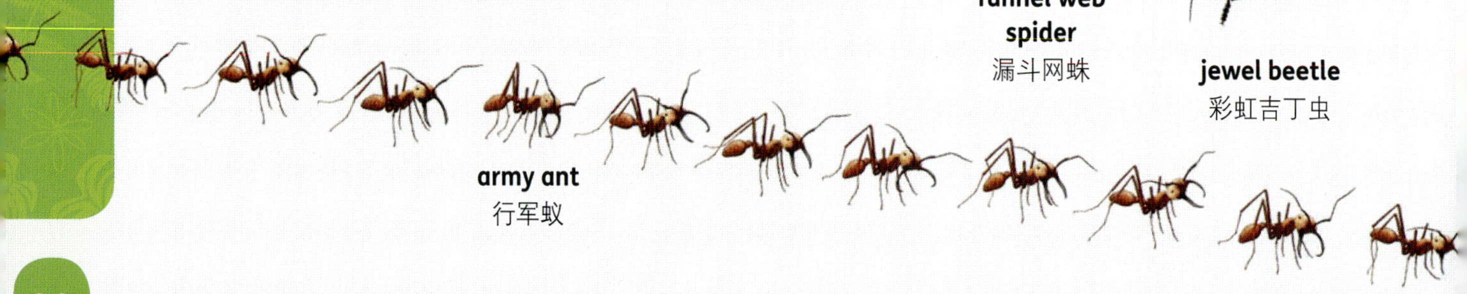

army ant
行军蚁

Rainforest creatures 雨林动物

vine snake 藤蛇

lowland gorilla 低地大猩猩

okapi 獾㹢狓（huò jiā pí）

African forest elephant 非洲森林象

kiwi 几维
Kiwis live in the temperate rainforests of New Zealand.
几维生活在新西兰的温带雨林。

leafcutter ant 切叶蚁

cassowary 鹤鸵

goliath spider 捕鸟蛛
The goliath spider is large enough to prey on small birds.
捕鸟蛛体形大，能捕杀小鸟。

Life on the rainforest floor 林地层动物

Rainforest creatures 雨林动物

Creatures of the canopy 树冠层动物

The rainforest canopy is a dense tangle of branches, leaves, fruit and flowers. It is home to more creatures than any other layer of the rainforest.

密集的树枝、树叶、果实和花朵构成了雨林的树冠层，在树冠层生活的动物比雨林其他各层都多。

orang-utan
猩猩

lemur 狐猴

rainbow lorikeet
彩虹吸蜜鹦鹉

eyelash viper
睫角棕榈蝮（fù）

honeyeater
吸蜜鸟

capuchin monkey
卷尾猴

giant rainforest praying mantis
澳洲斧螳
The praying mantis was given its name because it looks as though it is praying.
螳螂（táng láng）的姿势像在祷告。

tree kangaroo
树袋鼠

32

Rainforest creatures 雨林动物

hummingbird 蜂鸟
The hummingbird extends its long tongue to suck nectar from flowers.
蜂鸟伸出长舌头吸取花蜜。

colugo
鼯猴

tree snail
树蜗牛

flying gecko
褶虎

colobus monkey
疣（yóu）猴

tarsier
眼镜猴

Warning colours 警戒色

Some creatures are very brightly coloured. This dramatic colouring warns any predators that the creatures are poisonous. Poison dart frogs are often very small and the brighter their colouring, the more poisonous they are.

有些动物的颜色非常明艳，它们用醒目的颜色来警告捕食者它们有毒。箭毒蛙一般体形很小，它们的颜色越鲜艳，毒性就越强。

poison dart frogs 箭毒蛙
Poison dart frogs were given their name because the rainforest people dipped their hunting darts in the frogs' poison.
箭毒蛙之所以叫这个名字，是因为雨林的原住民会把吹箭的箭头浸入这种蛙的毒素中。

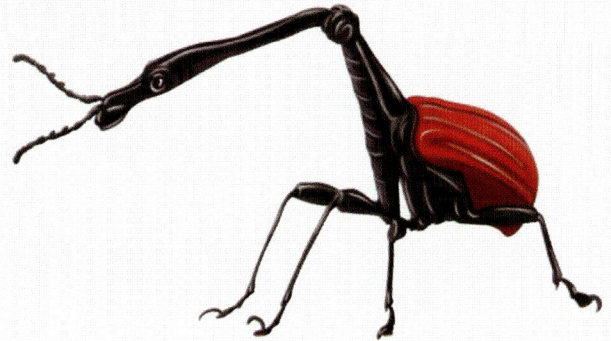

giraffe weevil 长颈象甲
The giraffe weevil uses its very long neck to help it fight.
长颈象甲用长脖子来打斗。

Creatures of the canopy 树冠层动物

Rainforest creatures 雨林动物

In the Amazon rainforest
在亚马孙雨林

The Amazon rainforest contains millions of animal species. The creatures shown here live in the Amazon canopy. Most canopy-dwellers are vegetarian, but some prey on small creatures.

亚马孙雨林中生活着上百万种动物。这两页图中的动物生活在亚马孙雨林的树冠层。大多数生活在树冠层的动物吃素，也有一部分以小动物为食。

❶ **scarlet macaw**
绯（fēi）红金刚鹦鹉

❷ **woolly monkey**
绒毛猴

❸ **Amazon parrot**
亚马孙鹦哥

❹ **morpho butterfly**
闪蝶

❺ **green iguana**
美洲鬣蜥

❻ **three-toed sloth**
三趾树懒
Sloths sleep for up to 20 hours a day.
树懒每天最多可以睡 20 小时。

❼ **spider monkey**
蜘蛛猴

❽ **coatimundi**
长鼻浣熊

❾ **keel-billed toucan**
厚嘴鵎鵼

❿ **pygmy marmoset**
倭狨（wō róng）

⓫ **howler monkey**
吼猴
Howler monkeys are known for the loud howls they make at the beginning and end of the day.
在每天清晨和黄昏，吼猴都会发出响亮的吼叫声，因此而得名。

⓬ **poison dart frog**
箭毒蛙

34

Rainforest creatures 雨林动物

In the Amazon rainforest 在亚马孙雨林

35

Beneath the shelter of the trees are countless creatures. Animals, birds and insects live on the forest floor or make their homes in tree trunks and branches...

在树林的隐蔽处生活着数不清的动物。兽类、鸟类、昆虫要么在林地层安家,要么住在树干里或树枝上……

FOREST AND WOODLAND WILDLIFE

森林和林地动物

Forest and woodland wildlife 森林和林地动物

Forest habitats 森林栖息地

There are two main types of forest in the cooler parts of the world. Deciduous forests have trees that shed their leaves in winter. They are sometimes known as broadleaf forests. Some forests have trees that stay green all year round. They are sometimes called evergreen forests.

在气温寒凉的地带主要有两种森林。落叶林的树木到了冬天树叶会掉落，有时又称为阔叶林。有些森林的树木一年四季都是绿的，有时称为常绿林。

Forest habitats are in danger all over the world. As the world population grows, people settle in regions that were once covered with forest, and trees are cut down for timber and fuel.

随着世界人口增长，人类开辟森林并定居下来，砍伐树木作为建造材料或燃料，全世界的森林栖息地都受到威胁。

Deciduous forests and woodlands
落叶森林和林地

Deciduous forests usually grow in regions where there is plenty of rain. There are many deciduous forests in Europe and eastern America.
落叶林往往生长在雨水充沛的地区，欧洲和美洲东部有大量的落叶林。

brimstone butterfly
硫磺（huáng）蝶

garden snail
花园蜗牛

woodpecker
啄木鸟

fallow deer
黇（tiān）鹿

chipmunk
花鼠

38

Forest and woodland wildlife 森林和林地动物

Major regions of deciduous and evergreen forest
主要落叶林和常绿林分布图

- **deciduous forest** 落叶林
- **evergreen forest** 常绿林

Evergreen forests 常绿林

Evergreen forests mostly grow in places with long, snowy winters and short, cool summers. Large parts of northern Canada and Russia are covered by evergreen forests.

常绿林生长的地方一般有漫长多雪的冬季和短暂凉爽的夏季。加拿大和俄罗斯的北部地区有大片常绿林。

hawk owl 鹰鸮（xiāo）

red squirrel 赤松鼠

pine marten 松貂（diāo）

wolf 狼

bobcat 短尾猫

Forest habitats 森林栖息地

39

Forest and woodland wildlife 森林和林地动物

Deciduous forest creatures 落叶林动物

Many animals in deciduous forests feed on nuts, berries and insects. When the trees lose their leaves, most mammals hibernate. They wake up again when spring arrives and there is plenty of food to eat.

许多住在落叶林的动物都以坚果、浆果和昆虫为食。树叶掉光时，大部分哺乳动物便会冬眠。直到春天来临，食物充足的时候，它们才会苏醒。

nightingale
夜莺
The nightingale is famous for its beautiful song.
夜莺以歌声优美而闻名。

wild turkey
野火鸡
Wild turkeys roam free in some American woodlands.
野火鸡在美洲的林地中游荡。

red deer
欧洲马鹿
Some creatures, like red deer and grey squirrels, live in both deciduous and evergreen forests.
有些动物在落叶林和常绿林都有，如欧洲马鹿和灰松鼠。

raccoon
浣熊

Forest and woodland wildlife 森林和林地动物

weasel
伶鼬 (yòu)

wood warbler
林莺

wood pigeon
林鸽

grey squirrel
灰松鼠

Tasmanian devil
袋獾 (huān)
The Tasmanian devil is only found in Tasmania, Australia.
袋獾现今只分布于澳大利亚的塔斯马尼亚。

red fox
赤狐

tawny owl
灰林鸮
The tawny owl is also known as the brown owl.
灰林鸮又叫 brown owl。

Deciduous forest creatures 落叶林动物

41

Forest and woodland wildlife 森林和林地动物

Life on a woodland floor
林地地面动物

This scene shows creatures in a deciduous woodland. Some dig burrows in the earth. Some live in rotting leaf mould on the forest floor. The minibeasts shown under the magnifier are all found in the leaf mould of a deciduous wood.

右边的场景图展现了落叶林地中的动物，有的在地上挖洞，有的住在地面的腐叶土中。放大镜下的小型无脊椎动物都生活在落叶树林的腐叶土里。

❶ stoat
白鼬

❷ mole
鼹（yǎn）鼠

❸ badger
獾

❹ dormouse
睡鼠

❺ shrew
鼩鼱（qú jīng）

❻ hedgehog
刺猬
Hedgehogs are protected by spiny quills.
刺猬用身上的尖刺来保护自己。

Creatures in the leaf mould
腐叶土里的动物

❼ woodlouse
鼠妇

❽ centipede
蜈蚣

❾ earthworm
蚯蚓

❿ millipede
马陆

Forest and woodland wildlife 森林和林地动物

Life on a woodland floor 林地地面动物

Forest and woodland wildlife 森林和林地动物

Evergreen forest wildlife 常绿林动物

Creatures in an evergreen forest need to cope with freezing, snowy winters. Some mammals grow thick coats and some hibernate. Birds usually migrate south for the winter. They return to the forests in the spring, when the weather warms up and they can find enough food to eat.

生活在常绿林的野生动物需要抵抗冰冷多雪的冬天。有的哺乳动物长出厚厚的皮毛，有的冬眠，鸟儿通常会迁徙到南方过冬。等到春天，天气转暖，食物充足，它们再回到林中。

Steller's jay
暗冠蓝鸦

black bear
黑熊

skunk
臭鼬
If they are attacked, skunks release a very strong smelling spray.
遭受攻击时，臭鼬会释放难闻的臭味。

lynx
猞猁（shē lì）

Forest and woodland wildlife 森林和林地动物

treecreeper
旋木雀

great grey owl
乌林鸮
Owls are nocturnal, so they usually hunt at night.
鸮是夜行动物，在夜间捕猎。

moose
驼鹿
The male moose uses its giant antlers to fight other males.
雄性驼鹿用巨大的鹿角相互打斗。

polecat
鸡鼬

wolverine
狼獾

porcupine
豪猪

Evergreen forest wildlife 常绿林动物

45

Eagles soar among the high mountains. Snow leopards stalk their prey on rocky peaks, and pandas live in forests on the lower slopes...

雕在高山上空翱翔，雪豹在崎岖的山峰追踪猎物，熊猫生活在山麓的森林中……

MOUNTAIN ANIMALS
山地动物

本章音频

Mountain animals 山地动物

Mountains habitat 山地栖息地

Mountains are found in all the world's continents. The Andes, the Rockies, the Himalayas and the Alps are all major mountain ranges. Some very high mountain peaks are covered with snow all year round. Only a few species can survive in these harsh conditions. Some mountains in tropical regions have dense rainforests on their lower slopes. Pandas and gorillas are found in these mountain forests.

世界各大洲都有山地。安第斯山、落基山、喜马拉雅山和阿尔卑斯山是主要的山脉。有些高峰终年被积雪覆盖，只有少数物种能在这些恶劣的环境中生存。在热带，有的山麓（lù）生长着茂密的雨林。熊猫和大猩猩就生活在这种山地森林中。

Upper and lower slopes 山顶和山麓

Most mountains provide two different habitats. The steep upper slopes are bare and rocky and often blanketed with snow. The gentle lower slopes are generally covered with trees.

大多数山地提供了两种栖息地。靠近山顶的斜坡较陡峭，植被稀少，岩石嶙峋，常有积雪。平缓的山麓一般树木繁茂。

Upper slopes
山顶斜坡

mountain goat
雪羊

bearded vulture
胡兀鹫（wù jiù）

Lower slopes
山麓

vicuña
骆马

red panda
小熊猫

Mountain animals 山地动物

Major mountain ranges 主要山脉分布图

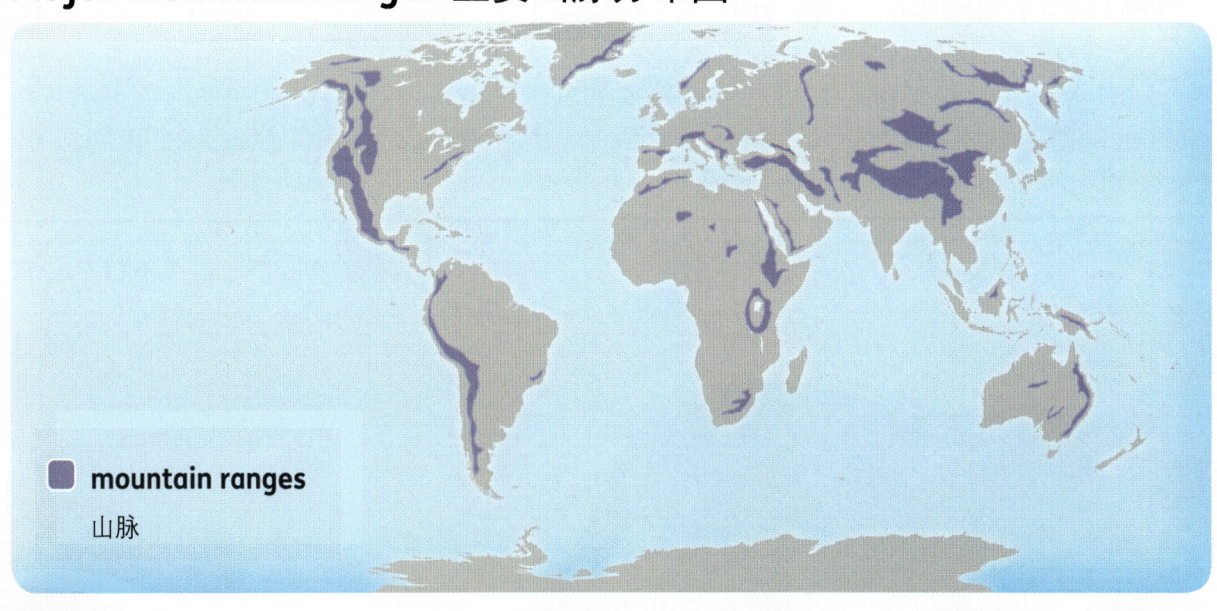

- mountain ranges
  山脉

snow leopard
雪豹

chinchilla
毛丝鼠

crimson rosella
红玫瑰鹦鹉

mountain gorilla
山地大猩猩

Mountain habitats 山地栖息地

Mountain animals 山地动物

Life in the mountains 山地动物

Finding food and shelter on the mountain tops is hard, so animals like the chamois and cougar move between the upper and lower slopes. Most other creatures stay in the highland forests. The giant panda lives only in mountain forests as its lowland habitat has been destroyed. It feeds on the bamboo plants that grow in the forests.

在山顶上很难找到食物和隐蔽处，因此臆（yì）羚和美洲狮等动物在山麓和山顶之间活动。其他动物大多生活在高地森林中，大熊猫只生活在山地森林，因为它们从前的低地栖息地已被破坏。大熊猫吃森林里的竹子。

chamois
臆羚

cougar
美洲狮
Puma is another name for cougar.
美洲狮还有个名字叫puma。

Mountain birds 山地鸟类

Many birds of prey hunt on mountain slopes where they can easily spot their prey. Their large, powerful wings allow them to soar to very high places. But birds of prey are not only found in mountain regions. They are also at home in other habitats, such as open grasslands, where they can view their prey.

许多猛禽在容易发现猎物的山坡上捕猎，强壮的大翅膀让它们能直冲云霄。猛禽不光山地有，也在其他栖息地定居，如开阔的草原，在那儿它们能看清猎物。

red kite
赤鸢（yuān）

common buzzard
普通鵟（kuáng）

peregrine falcon
游隼（sǔn）

golden eagle
金雕

50

Mountain animals 山地动物

Domesticated creatures 驯化的动物

Some mountain creatures have been domesticated by humans. They have been bred as working animals, pets and farm animals. Yaks and llamas are used for carrying loads, and for their wool and meat. Alpacas are bred for their very fine wool.

有些山地动物已被人类驯化，培育为役畜、宠物或农畜。牦牛和家羊驼被用来运货，它们的肉和毛也有用处。人类驯养小羊驼，以剪取它们细密的皮毛。

alpaca
小羊驼

yak
牦（máo）牛

llama
家羊驼

cavy
豚鼠
Cavies are the wild ancestors of guinea pigs.
豚鼠是天竺（zhú）鼠的野生原始种。

giant panda
大熊猫
Giant pandas are in danger of becoming extinct.
大熊猫濒临灭绝。

ibex
羱（yuán）羊

Life in the mountains 山地动物

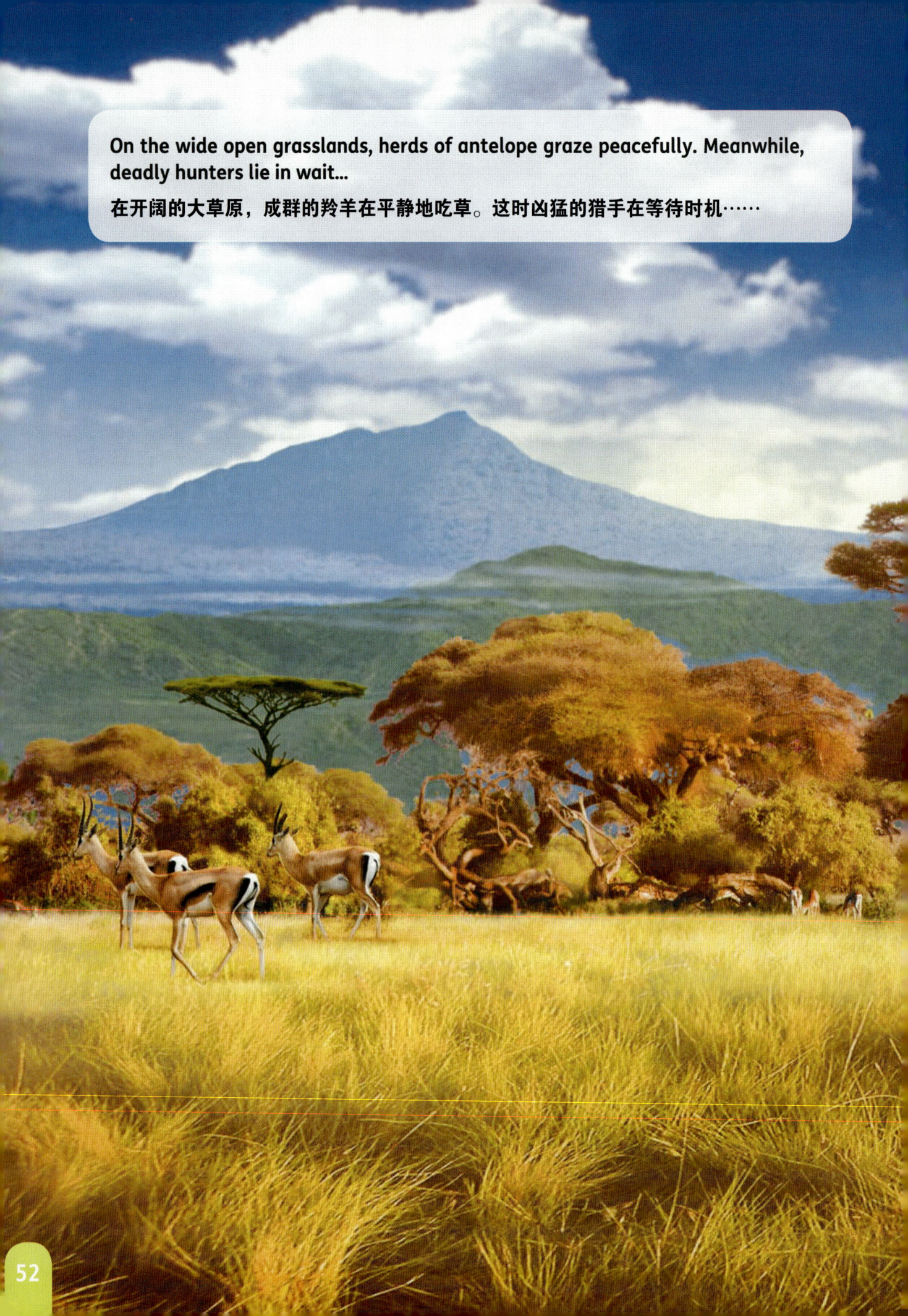

On the wide open grasslands, herds of antelope graze peacefully. Meanwhile, deadly hunters lie in wait...

在开阔的大草原，成群的羚羊在平静地吃草。这时凶猛的猎手在等待时机……

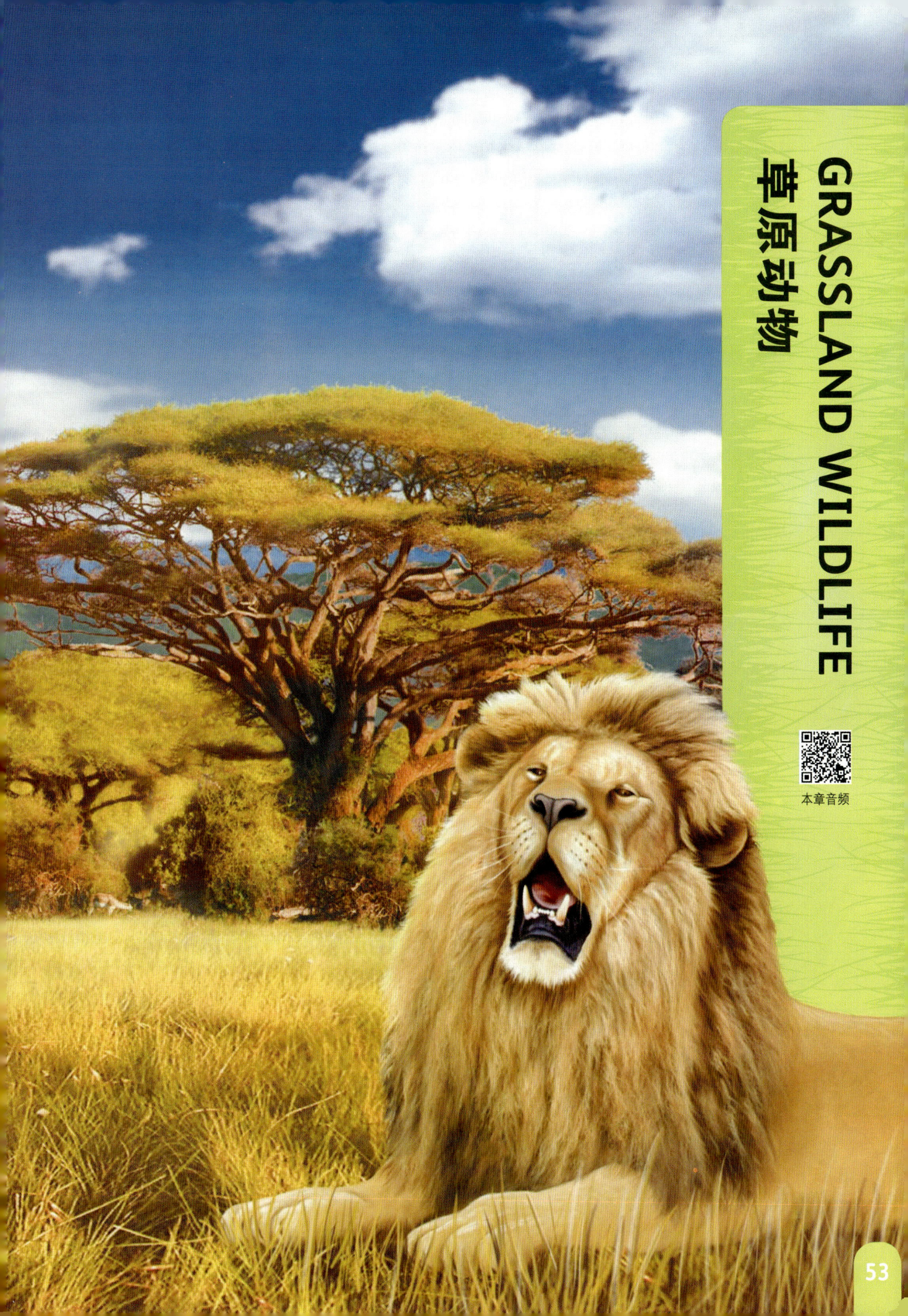

GRASSLAND WILDLIFE

草原动物

Grassland wildlife 草原动物

Grassland habitats 草原栖息地

Grasslands are large stretches of ground covered by wild grasses and shrubs. They are found in both hot and cold regions, and are given different names in different parts of the world.

草原是长满野草和灌木丛的广阔陆地。草原在炎热地区和寒冷地区都有，世界不同地方的草原有不同的名称。

Hot, dry grasslands 炎热干燥的草原

Some hot, dry grasslands are called savannahs. There are savannahs in Africa, Asia and South America. The South American grasslands are sometimes known as the Pampas.

有些炎热干燥的草原被称为稀树草原或萨瓦纳草原。非洲、亚洲和南美洲都有稀树草原。南美洲的草原有时又被称为潘帕斯草原。

ostrich 鸵鸟

gazelle 瞪羚

Cooler grasslands 气候凉爽的草原

The grasslands of North America are called the Prairies. They have hot summers and cold, wet winters. There are also cool grasslands in northern Asia. They are known as the Steppes.

北美洲的草原叫作北美大草原。大草原上夏天炎热，冬天寒冷潮湿。亚洲北部也有凉爽的草原，称为温带草原。

bison 野牛

jackrabbit 杰克兔

54

Grassland wildlife 草原动物

Major grassland regions 主要草原分布图

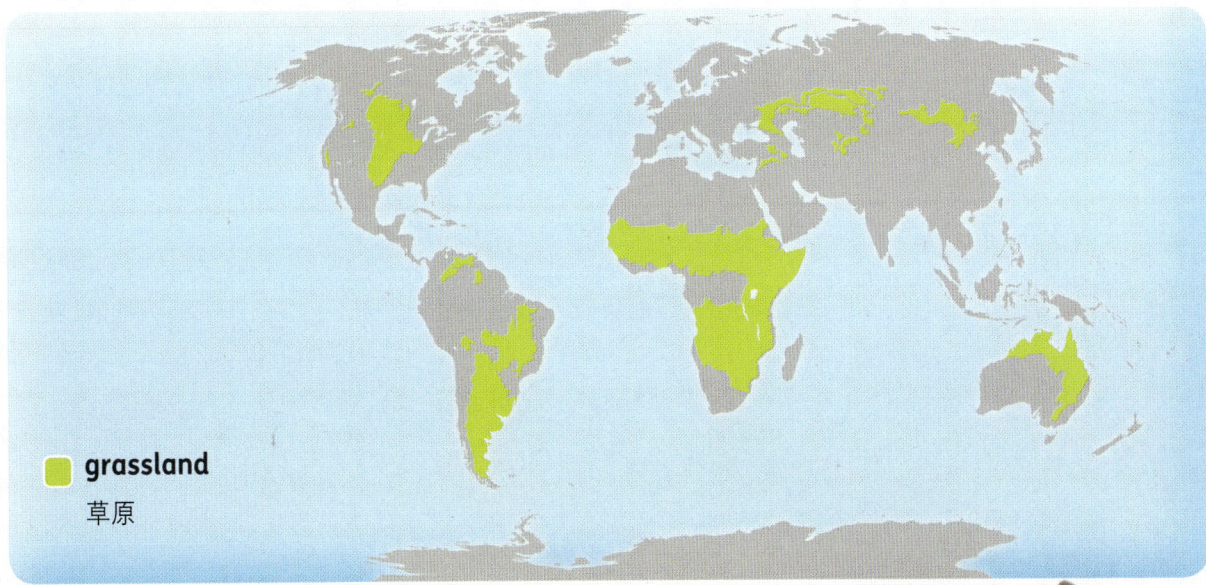

grassland
草原

Bush and scrub

荒野和灌木丛

Many parts of Australia are covered with grasses and scrub. These hot, dry regions are known as the bush. Eucalyptus trees, or gum trees, also grow in many parts of the Australian bush.

澳大利亚大部分地区被草地和灌木丛覆盖。这些炎热干燥的地区称为荒野。在澳大利亚荒野上还生长着桉（ān）树（又叫树胶树）。

wombat
毛鼻袋熊

goanna
澳洲巨蜥

common adder
极北蝰

pheasant
环颈雉（zhì）

Moorlands 高沼地

Moors are found in cool regions, with plenty of rain. They have low bushes and shrubs which provide good cover for wildlife. In winter, they are often covered in snow. In the warmer months, they can be boggy.

高沼地分布在凉爽、多雨的地区。高沼地上有低矮的灌木丛，为野生动物提供了良好的隐蔽场所。在冬季，高沼地通常被积雪覆盖。到了温暖的月份，高沼地会变得松软泥泞。

Grassland habitats 草原栖息地

Grassland wildlife 草原动物

On the savannah 在稀树草原上

On the African savannah, many animals live in herds. They travel large distances in search of water, and they are often in danger from hunters.

在非洲稀树草原，许多动物都成群结队行动。它们长途跋涉寻找水源，常有被猎手盯上的危险。

❶ lion
狮子

❷ buffalo
水牛

❸ zebra
斑马

Grassland wildlife 草原动物

On the savannah 在稀树草原上

4 giraffe
长颈鹿

5 cheetah
猎豹

6 African elephant
非洲象

7 antelope
羚羊

57

Grassland wildlife 草原动物

More savannah creatures 稀树草原上的另一些动物

Many animals of the savannah escape from predators by running very fast. Some burrow underground to stay safe. The rhinoceros and the armadillo are protected from attack by their tough outer covering.

稀树草原上的许多动物跑得非常快，以逃避食肉动物追捕。有的动物挖地洞来躲避天敌。犀牛和犰狳（qiú yú）有坚硬的铠甲保护自己。

aardvark
土豚
Aardvarks eat ants and termites.
土豚吃蚂蚁和白蚁。

meerkat
细尾獴（měng）

mongoose
獴

rhinoceros
犀牛

royal python
球蟒

Grassland wildlife 草原动物

wildebeest
角马

armadillo
犰狳
The armadillo's body is protected by bony plates.
犰狳的身体由骨板保护。

warthog
疣猪

Termite mound 白蚁丘

Termites live in colonies with a single large queen. They build a mound to house their queen, and she lays eggs which will hatch to form all the young in the colony. Some termite mounds are as tall as a two-storey house!

白蚁成群聚居，蚁群中有一只巨大的蚁后。白蚁修筑蚁丘供蚁后居住，蚁后产卵，孵出幼虫。有的白蚁丘甚至有两层楼高！

ventilation shaft
中央通气道

chimney
侧通气道

nursery galleries for newly hatched young
幼虫孵化室

royal cell for the queen
蚁后巢穴

termite
白蚁
Termites are sometimes called white ants.
白蚁有时又叫蟗（wèi）。

cutaway termite mound
白蚁丘剖面图
Termite workers build mounds from earth and saliva. Soldier termites defend it from attack.
工蚁用泥土和唾液修筑蚁丘，兵蚁保护蚁丘免遭攻击。

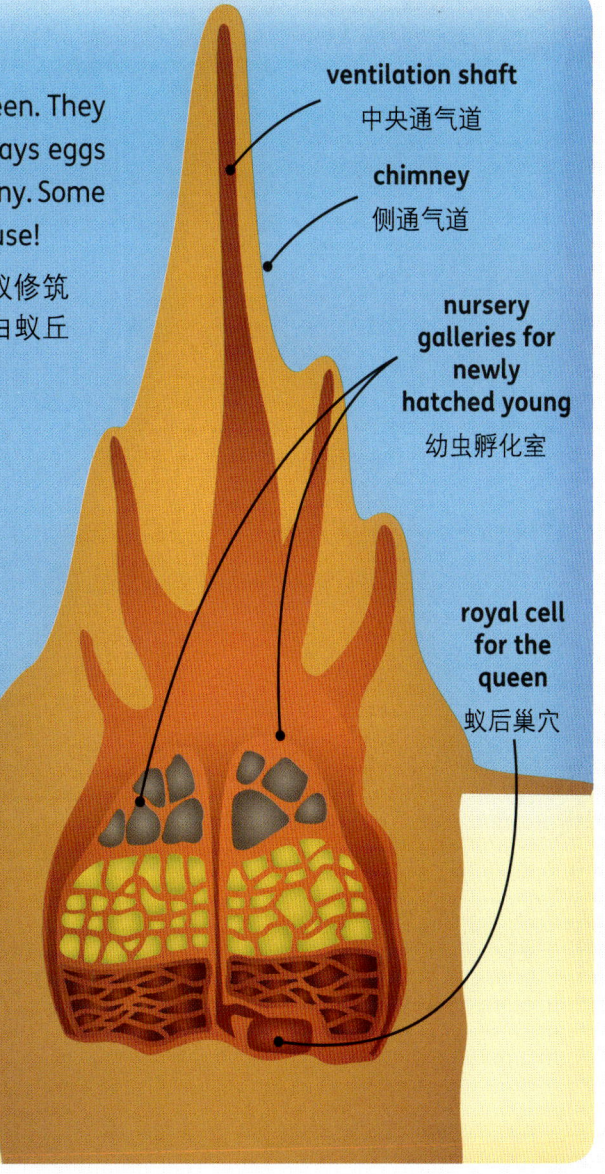

More savannah creatures 稀树草原上的另一些动物

59

Grassland wildlife 草原动物

Grassland and moorland wildlife 草原和高沼地动物

Grassland and moorland animals are usually very well camouflaged. This helps to keep them safe from attack. Some moorland birds, such as grouse and partridges, fly close to the ground using bushes as cover.

草原和高沼地动物通常都很善于伪装，使自己免遭攻击。有些高沼地的鸟类，如松鸡和山鹑（chún），贴着地面低飞，以灌木丛作为掩护。

emperor moth
帝王蛾

guinea fowl
珍珠鸡

coyote
郊狼

Coyotes live on the North American prairies.
郊狼生活在北美大草原。

red grouse
柳雷鸟

Grassland wildlife 草原动物

skylark 云雀

kestrel 红隼

hare 野兔

grasshopper 蚂蚱 (mà zha)

partridge 山鹑

prairie dog 草原犬鼠
The prairie dog is a kind of ground squirrel.
草原犬鼠是一种地松鼠。

ferret 雪貂

Grassland and moorland wildlife 草原和高沼地动物

61

Grassland wildlife 草原动物

In the bush 在荒野中

In the Australian bush, there are rough grasses, scrubby bushes and scattered eucalyptus trees. The bush is home to many creatures that are not found anywhere else in the world. Some of these creatures are marsupials (animals that carry their young in a pouch).

在澳大利亚荒野中，有参差的野草、低矮的灌木丛和零星分布的桉树。这里住着许多其他地方没有的动物，当中包括有袋类动物（把幼崽装在育儿袋中的动物）。

❶ red kangaroo 赤大袋鼠

❷ dingo 澳洲野犬

❸ blue-tongued skink 蓝舌蜥

❹ spiny anteater 针鼹

❺ wallaby 沙袋鼠
Wallabies look like small kangaroos.
沙袋鼠长得像小型的袋鼠。

❻ galah 粉红凤头鹦鹉

Grassland wildlife 草原动物

In the bush 在荒野中

7 bilby
兔耳袋狸

8 emu
鸸鹋（ér miáo）
Emus cannot fly but they can run very fast.
鸸鹋不会飞，但跑得非常快。

9 cockatoo
凤头鹦鹉

10 bandicoot
袋狸

11 budgerigar
虎皮鹦鹉

12 koala
树袋熊

63

Alligators lurk in shady swamps, waiting to snap at their prey. The waters and shores of the swamps are teeming with hidden creatures...

短吻鳄潜伏在阴暗的沼泽里,伺机咬住猎物。沼泽的水中和岸边隐藏着许多动物……

RIVER, LAKE AND WETLAND WILDLIFE
河流、湖泊和湿地的动物

本章音频

River, lake and wetland wildlife 河流、湖泊和湿地的动物

Water habitats 水域栖息地

Rivers, lakes, ponds and wetlands provide a range of habitats for water-dwelling creatures. They are also home to land-dwelling animals that live on the water's edge. Some species can only be found in a certain type of habitat, such as a mangrove swamp. Others live in a range of watery habitats.

河流、湖泊、池塘和湿地为水生动物提供了各种栖息地。住在水边的陆生动物也在这些地方安家。有的动物只在某些特定的栖息地生活，如红树林沼泽。有的动物在许多水域栖息地都能看到。

Rivers and riverbanks 河流与河岸

muskrat 麝（shè）鼠

hippopotamus 河马

rainbow trout 虹鳟（zūn）

Lakes and ponds 湖泊和池塘

catfish 鲇（nián）鱼

damselfly 豆娘

coot 骨顶鸡

66

River, lake and wetland wildlife 河流、湖泊和湿地的动物

Major rivers 主要河流分布图

■ major rivers
主要河流

Wetlands 湿地

Swamps, marshes and bogs are all types of wetland. There are freshwater and saltwater wetlands. Mangrove swamps are saltwater wetlands. They are found close to the coast in warm, wet regions of the world.

木本沼泽、草本沼泽和藓（xiǎn）类泥炭沼泽都是湿地。湿地又分为淡水湿地和咸水湿地。红树林沼泽是咸水湿地，分布在温暖潮湿的海岸边。

spoonbill
琵鹭（pí lù）

water snail
田螺

spectacled caiman
眼镜凯门鳄

Water habitats 水域栖息地

67

River, lake and wetland wildlife 河流、湖泊和湿地的动物

River creatures 河流动物

Rivers can be warm or icy, shallow or deep, fast or slow-moving. They also change their character, starting as a small, fast-flowing stream, and ending in a wide, shallow estuary. Each type of river is home to a different range of wildlife.

河流有的温暖，有的冰凉，有的深，有的浅，有的湍急，有的平缓。河流的特征也在变化，起初是一条快速流动的小溪，最后成为开阔的、浅浅的河口。每种河流里都生活着许多种野生动物。

crayfish
淡水螯虾

pike
狗鱼

mink
水貂

anaconda
水蚺
Anacondas live near rivers in the rainforest.
水蚺生活在雨林中的河边。

neon tetra
霓（ní）虹脂鲤
Neon tetras live in the Amazon River, but are often kept in an aquarium.
霓虹脂鲤生活在亚马孙河，如今常被养在鱼缸里。

crocodile
鳄鱼

River, lake and wetland wildlife 河流、湖泊和湿地的动物

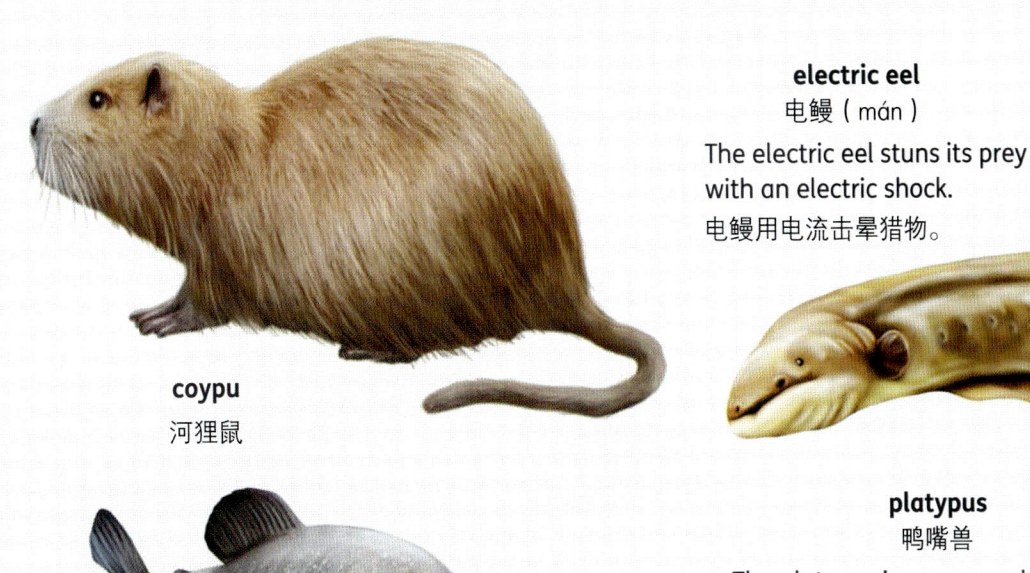

coypu
河狸鼠

electric eel
电鳗（mán）
The electric eel stuns its prey with an electric shock.
电鳗用电流击晕猎物。

piranha
食人鲳（chāng）
Piranhas use their very sharp teeth to strip the flesh off other creatures.
食人鲳用它们锋利的牙齿咬掉其他动物的肉。

platypus
鸭嘴兽
The platypus is a mammal that lays eggs. It is only found in Australia.
鸭嘴兽是产卵的哺乳动物。目前只在澳大利亚发现。

River creatures 河流动物

Beaver lodge 河狸巢穴

Beavers have a large, flat tail that they use to help them swim. They gather twigs to build a home called a lodge.
河狸有一条宽大扁平的尾巴帮它们在水中游动，它们收集树枝来修筑巢穴。

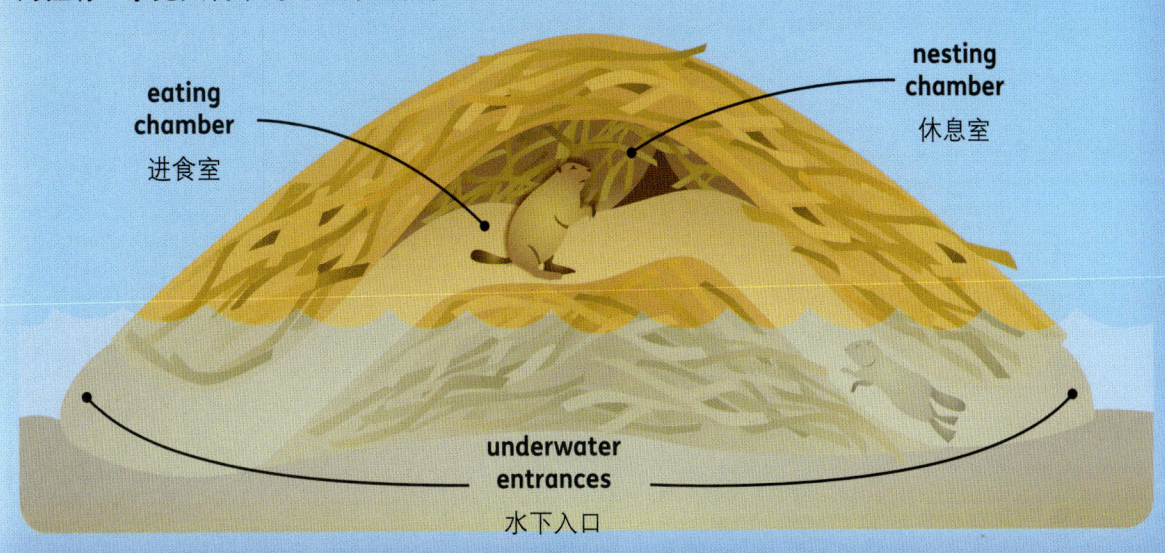

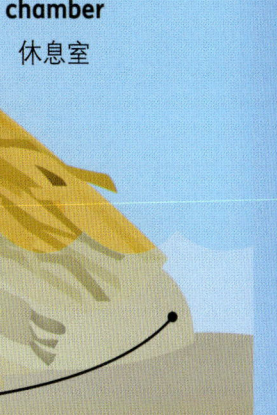

eating chamber
进食室

nesting chamber
休息室

underwater entrances
水下入口

69

River, lake and wetland wildlife 河流、湖泊和湿地的动物

Life on the river 河上的动物

Rivers provide a habitat for many creatures. Mammals live in burrows in the river bank. Ducks and swans glide over the water's surface, and birds catch fish and insects. Some insects fly very close to the water and some even walk on its surface!

河流为许多动物提供了栖息地。哺乳动物住在河岸上的洞穴，鸭子和天鹅划过水面，鸟儿捕捉鱼和昆虫。有些昆虫贴着水面低飞，有些甚至能在水面上行走！

1 heron 鹭

2 dipper 河乌
Dippers feed on insects.
河乌捕食昆虫。

3 water vole 水䶄（píng）

4 water shrew 水駒鼱

5 moorhen 黑水鸡

6 swan 天鹅

7 kingfisher 翠鸟
Kingfishers dive underwater to catch fish.
翠鸟潜入水中捕鱼。

8 otter 水獭（tǎ）

9 mallard 绿头鸭
Male mallards have brightly coloured plumage.
雄性绿头鸭有色彩明艳的羽毛。

10 common frog 黑斑蛙

Insects and minibeasts
昆虫和其他无脊椎动物

11 pond skater 水黾（mǐn）

12 dragonfly 蜻蜓

13 greater water boatman 松藻虫

14 mayfly 蜉蝣（fú yóu）

River, lake and wetland wildlife 河流、湖泊和湿地的动物

Life on the river 河上的动物

River, lake and wetland wildlife 河流、湖泊和湿地的动物

Lake and pond wildlife 湖泊和池塘动物

Most lakes and ponds are freshwater environments, although there are a few saltwater lakes. Few creatures live around salt lakes because these habitats provide little food or shelter.

虽然有些湖是咸水的，大多数湖泊和池塘是淡水环境。在咸水湖附近生活的动物很少，因为这些地方提供的食物和隐蔽场所太少。

snapping turtle
蛇鳄龟

American bullfrog
美洲牛蛙

pelican
鹈鹕（tí hú）

common toad
普通蟾蜍
Toads are often found close to ponds.
蟾蜍常在池塘附近出没。

flamingo
红鹳

capybara
水豚

72

River, lake and wetland wildlife 河流、湖泊和湿地的动物

basilisk lizard
双嵴（jí）冠蜥
Basilisk lizards can run on the surface of water.
双嵴冠蜥能在水面上飞奔。

osprey
鹗（è）

minnow
鲦（tiáo）

newt
水栖蝾螈

Carp 鲤鱼

The common carp is found in lakes. Goldfish and koi are domesticated species of carp. There are many different varieties of goldfish and koi.

鲤鱼生活在湖泊中，金鱼和锦鲤都是被驯化的鲤鱼种类，也分很多不同品种。

common carp
普通鲤鱼

goldfish
金鱼

koi
锦鲤

Lake and pond wildlife 湖泊和池塘动物

73

River, lake and wetland wildlife 河流、湖泊和湿地的动物

Wetland animals 湿地动物

Regions that are often flooded and waterlogged are known as wetlands. They include swamps, bogs, marshes, fens, reed beds and the land around river estuaries. A swamp is a wetland region with trees. Other types of wetland have very few trees.

湿地是经常被水淹没浸透的地带，包括木本沼泽、藓类泥炭沼泽、草本沼泽、草本泥炭沼泽、苇地、河口周围的陆地等。木本沼泽是有树木生长的湿地。其他几类湿地树木稀少。

ibis
鹮（huán）

crane
鹤

lungfish
肺鱼

Lungfish have lungs as well as gills so they can breathe air during a dry season.
肺鱼既有鳃又有肺，所以能在旱季呼吸空气。

alligator
短吻鳄

River, lake and wetland wildlife 河流、湖泊和湿地的动物

diamondback terrapin
钻纹龟

stork
白鹳

Wetland animals 湿地动物

Big cats of the swamp
沼泽里的大型猫科动物

Some big cats live in mangrove swamps, but they are in danger of becoming extinct because of poaching. Today, there are fewer than 200 Florida panthers in the wild.

有些大型猫科动物生活在红树林沼泽里，它们被偷猎而濒临灭绝。如今野生的佛罗里达美洲狮只剩不到 200 只。

Florida panther
佛罗里达美洲狮

Bengal tiger
孟加拉虎

75

In the desert, sand dunes stretch for miles. Scorpions, lizards and snakes bask in the baking sun...
荒漠里沙丘连绵。蝎子、蜥蜴和蛇在烈日下取暖……

DESERT WILDLIFE
荒漠动物

本章音频

Desert wildlife 荒漠动物

Desert habitats 荒漠栖息地

Deserts are vast regions of arid land with almost no rainfall. Some are hot and sandy. Others are cold and rocky. The world's largest hot desert is the Sahara in North Africa. The Atacama Desert in South America is a cold desert. It is one of the driest regions on Earth.

荒漠是广袤的干旱土地，降雨极少。有些荒漠炎热，被沙覆盖。有的寒冷，岩石遍布。世界上最大的热荒漠是北非的撒哈拉沙漠。南美洲的阿塔卡马沙漠是冷荒漠，是地球上最干燥的地区之一。

Hot deserts 热荒漠

Hot deserts have scorching days and cold nights. Powerful winds race over the dunes and whip up sandstorms. There are very few plants, except at an oasis. A water pool in a desert is called an oasis.

热荒漠白天炎热，夜晚寒冷。强劲的风刮过沙丘，形成沙尘暴。荒漠里的植被稀少，只有绿洲里植被较多。绿洲是荒漠中的水塘。

monitor lizard
巨蜥

fennec fox
聊（guō）狐

scorpion
蝎子

Desert wildlife 荒漠动物

Main desert regions 主要荒漠分布图

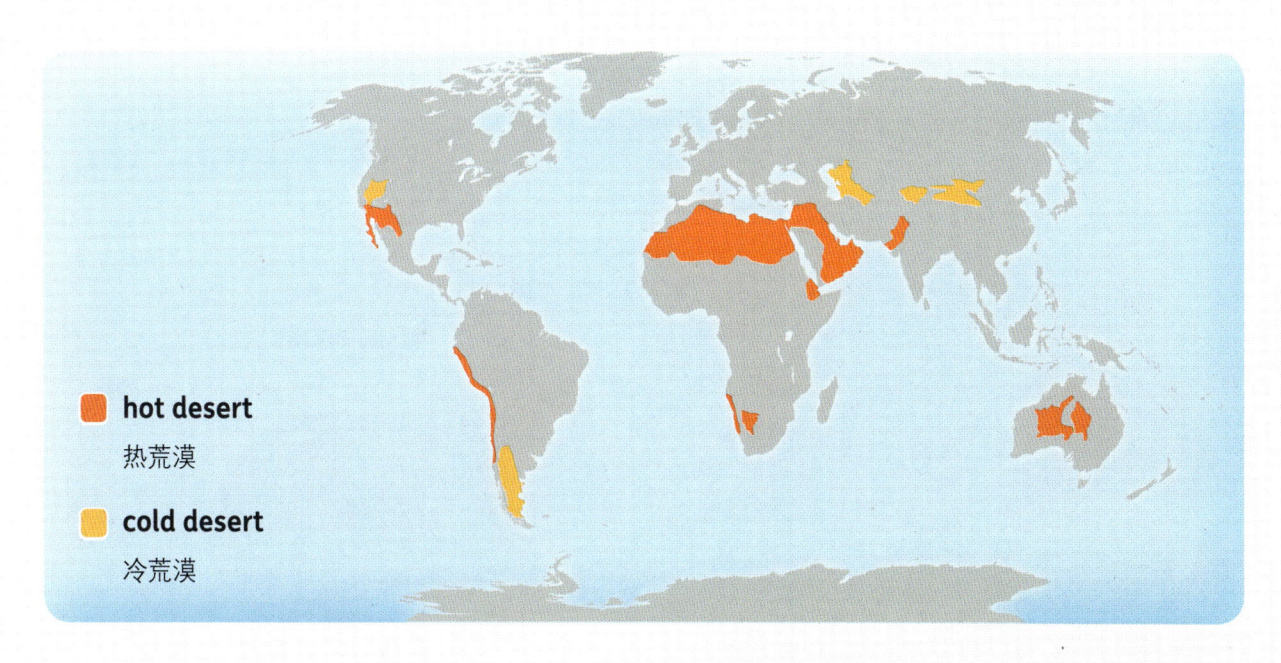

- hot desert 热荒漠
- cold desert 冷荒漠

Cold deserts 冷荒漠

Cold deserts can be warm during the day, but temperatures at night drop well below freezing point. Cactus plants often grow in cold deserts.

冷荒漠里白天可能很温暖,但是到了夜晚,温度会降到冰点以下。仙人掌类植物一般生长在冷荒漠中。

desert tortoise 阿氏沙龟

Bactrian camel 双峰驼

sidewinder rattlesnake 角响尾蛇

Desert habitats 荒漠栖息地

Desert wildlife 荒漠动物

Desert creatures 荒漠动物

Desert animals have to survive in a very harsh environment with almost no water. Some are able to store liquid in their bodies. Some burrow under the sand to keep cool. Many desert mammals are nocturnal. They only come out at night when it is cooler.

荒漠动物必须在极度缺水的严酷环境中生存。有的动物能在身体里储存液体。有的挖洞躲在沙里以保持凉爽。许多荒漠哺乳动物都是夜行动物，它们只在凉爽的夜晚行动。

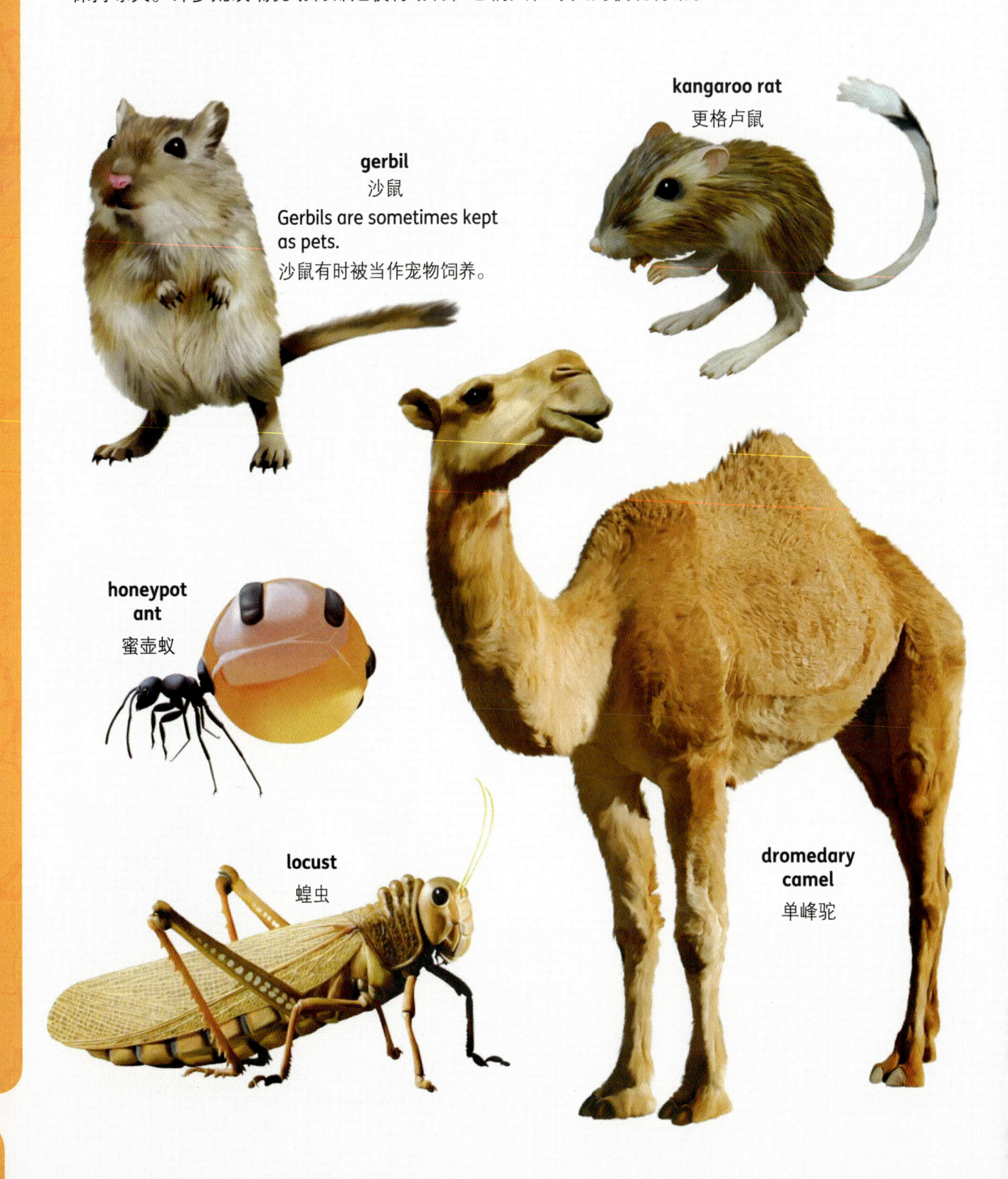

kangaroo rat 更格卢鼠

gerbil 沙鼠
Gerbils are sometimes kept as pets.
沙鼠有时被当作宠物饲养。

honeypot ant 蜜壶蚁

locust 蝗虫

dromedary camel 单峰驼

Desert wildlife 荒漠动物

black widow spider
黑寡妇蜘蛛

turkey vulture
红头美洲鹫
Vultures feed on the flesh of dead animals, known as carrion.
鹫的食物是动物尸体上的肉，也就是腐肉。

jerboa
跳鼠
The enormous ears of the jerboa allow it to lose body heat rapidly.
跳鼠的大耳朵帮助它们快速散热。

horned desert viper
角蝰

roadrunner
走鹃

inland taipan
内陆太攀蛇
The inland taipan is one of the world's most venomous snakes.
内陆太攀蛇是毒性极强的蛇。

Desert creatures 荒漠动物

81

In the depths of the ocean, giant manta rays glide slowly and silently...
在海洋深处，巨蝠鲼（fèn）缓慢无声地游过……

LIFE IN THE OCEAN
海洋动物

本章音频

Oceans 海洋

Life in the ocean 海洋动物

Oceans 海洋

More than two-thirds of the Earth's surface is covered by water. There are five oceans and several seas and they are home to a vast range of creatures. Creatures that live in the oceans include mammals, such as whales and dolphins, fish of all sizes, crustaceans, like lobsters and shrimps, and very simple life forms, such as tube worms.

地球表面超过三分之二的区域被水覆盖。地球上总共有五大洋和好几片大海，海洋中居住着种类繁多的动物。生活在海洋中的动物有哺乳动物，如鲸、海豚，体形各异的鱼类，还有甲壳动物，如龙虾和小虾，以及低等的动物，如管虫。

Ocean zones 海洋区域

Oceans and seas can be divided into five different zones. Each zone has its own particular wildlife, but some creatures move between different zones.

海洋可以划分为五个不同的区域。每个区域都有其独特的野生动物，有的动物在不同的区域间活动。

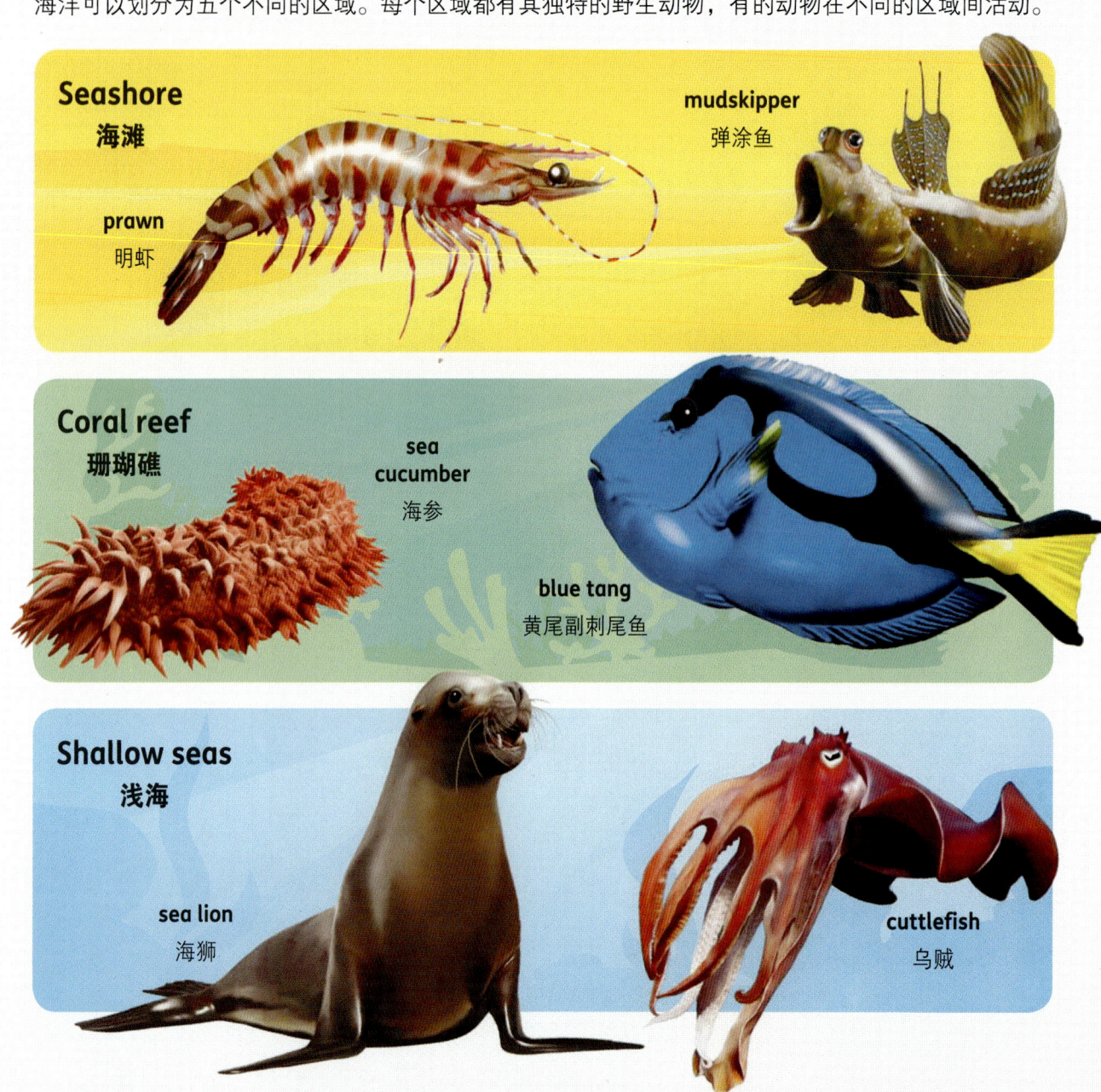

Seashore 海滩
- prawn 明虾
- mudskipper 弹涂鱼

Coral reef 珊瑚礁
- sea cucumber 海参
- blue tang 黄尾副刺尾鱼

Shallow seas 浅海
- sea lion 海狮
- cuttlefish 乌贼

84

Life in the ocean 海洋动物

Major oceans and seas 主要海洋分布图

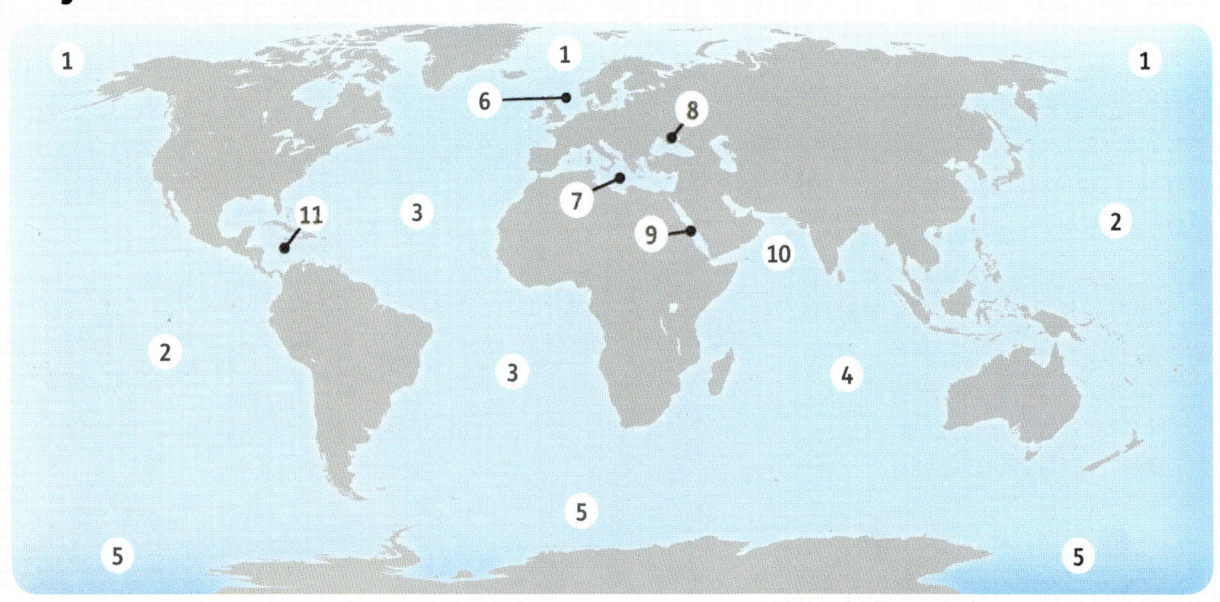

1. Arctic Ocean 北冰洋
2. Pacific Ocean 太平洋
3. Atlantic Ocean 大西洋
4. Indian Ocean 印度洋
5. Southern Ocean 南大洋
6. North Sea 北海
7. Mediterranean Sea 地中海
8. Black Sea 黑海
9. Red Sea 红海
10. Arabian Sea 阿拉伯海
11. Caribbean Sea 加勒比海

Open ocean 外海

herring 鲱 (fēi)

flying fish 飞鱼

Deep seas 深海

vampire squid 幽灵蛸 (shāo)

wolf fish 狼鳚 (wèi)

Oceans 海洋

85

Life in the ocean 海洋动物

On the seashore 在海滩上

Seashore creatures live on sand and rocks that are covered by the sea each time the tide comes in. Rock pools are home to many colourful creatures and seabirds fly along the shore looking for food.

海滩上的动物住在沙滩上、碎石间，每当涨潮时这片区域被海水淹没。岩池里住着许多色彩斑斓的小动物，海鸟沿着海滩低飞，寻找食物。

1 oystercatcher 蛎鹬（lì yù）

2 cormorant 鸬鹚（lú cí）

3 crab 螃蟹
Crabs scuttle sideways across the sand.
螃蟹在沙滩上快步横行。

4 shrimp 小虾

5 starfish 海星

6 sea urchin 海胆

7 mussel 贻贝
Mussels and barnacles feed on microscopic plankton in the water.
贻贝和藤壶以水中微小的浮游生物为食。

8 sea anemone 海葵
Sea anemones open up under water.
海葵在水中张开。

9 hermit crab 寄居蟹

10 limpet 帽贝

11 barnacle 藤壶

12 puffin 海鹦

13 seagull 海鸥

86

Life in the ocean 海洋动物

On the seashore 在海滩上

Life in the ocean 海洋动物

In the ocean 在海洋中

Thousands of species of fish live in the oceans. Smaller fish usually swim in groups, called shoals, while larger fish hunt alone. The fish on these two pages are shown roughly to scale.

海洋中生活着成千上万种鱼类。体形较小的鱼一般聚在一起游动，称为鱼群，体形大的鱼独自捕猎。这两页图中的鱼是按大致比例绘制的。

① sardine 沙丁鱼
② mackerel 鲭（qīng）
③ plaice 鲽（dié）
④ pilot fish 舟鲕（shī）
⑤ flounder 偏口鱼
⑥ barracuda 大舒（yú）
⑦ halibut 大比目鱼
⑧ sole 鳎（tǎ）
⑨ dogfish 狗鲨

Life in the ocean 海洋动物

In the ocean 在海洋中

10 monkfish 鮟鱇（ān kāng）
11 sea bass 海鲈
12 marlin 马林鱼
13 cod 鳕（xuě）
14 haddock 黑线鳕
15 tuna 金枪鱼
16 swordfish 剑鱼

89

Life in the ocean 海洋动物

More ocean life 另一些海洋动物

The ocean is home to some very large creatures. Whales, dolphins, porpoises, manatees and dugongs are all mammals. Sharks and rays are fish. Sea mammals like whales need air to breathe. They swim up to the surface to blow out used air and breathe in fresh air.

海洋中有一些庞大的动物。鲸、海豚、鼠海豚、海牛和儒艮（gèn）是哺乳动物。鲨鱼和鳐（yáo）是鱼类。鲸等海洋哺乳动物都需要呼吸空气。它们浮上海面喷出废气，吸入新鲜空气。

dolphin
海豚
Dolphins use sounds, such as clicks and whistles, to communicate with each other.
海豚通过叫声相互交流，如吱吱声和口哨声。

manatee
海牛

hammerhead shark
双髻（jì）鲨

manta ray
蝠鲼

great white shark
大白鲨
A great white shark can live to be as old as 70.
大白鲨能活到 70 岁。

dugong
儒艮
Dugongs graze on seagrasses on shallow sea beds.
儒艮吃长在浅海海床的海草。

porpoise
鼠海豚
Whales, dolphins and porpoises breathe through a blowhole.
鲸、海豚和鼠海豚通过呼吸孔呼吸。

Life in the ocean 海洋动物

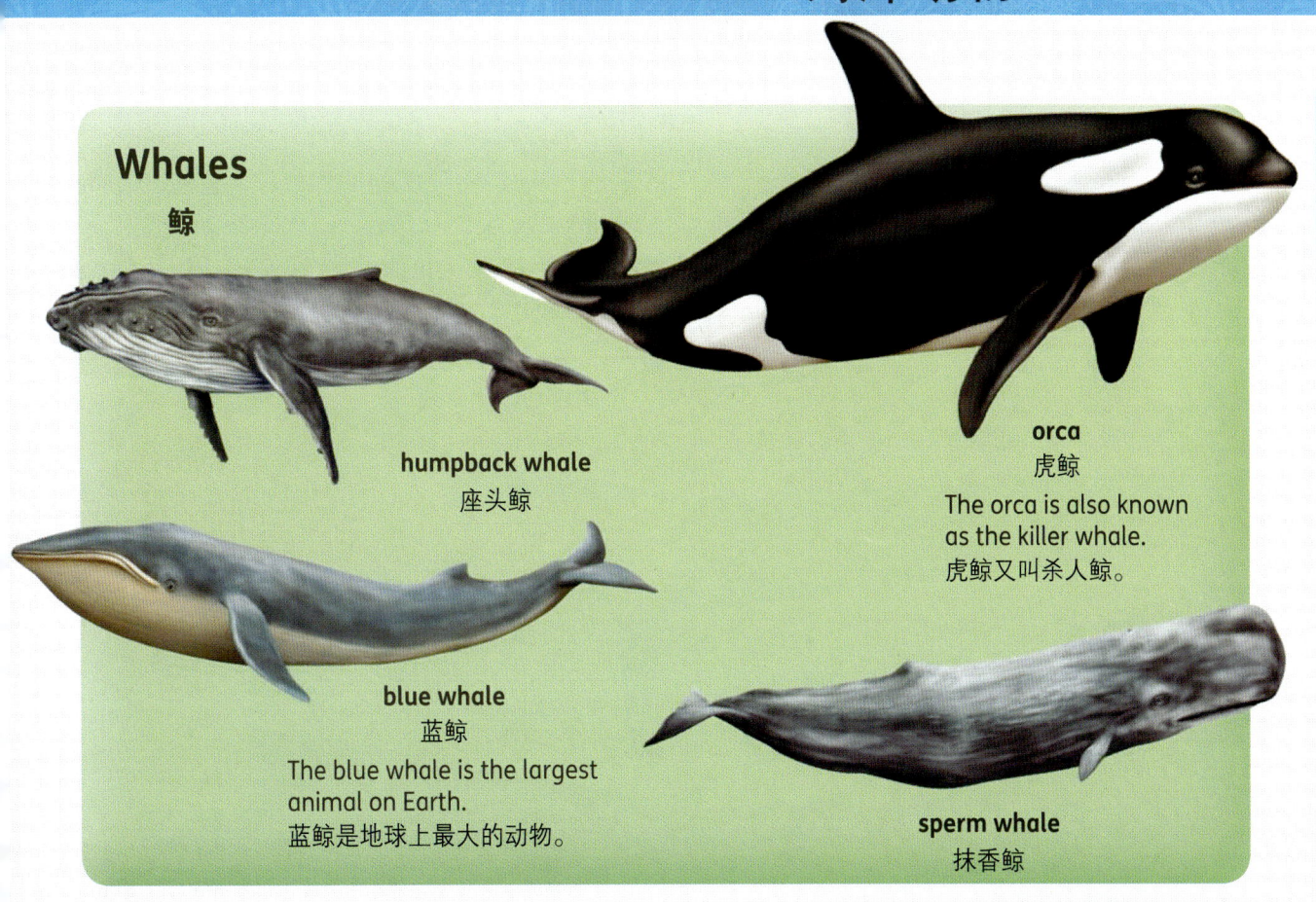

Whales 鲸

humpback whale
座头鲸

orca
虎鲸
The orca is also known as the killer whale.
虎鲸又叫杀人鲸。

blue whale
蓝鲸
The blue whale is the largest animal on Earth.
蓝鲸是地球上最大的动物。

sperm whale
抹香鲸

Baleen whales 须鲸

Some whales have flexible bristles called baleen inside their mouths to help them feed. The baleen plates trap tiny sea creatures, such as plankton and krill. When a whale raises its tongue the plankton and krill are trapped inside its mouth.

有些鲸的嘴里有柔韧的刚毛，叫作鲸须，帮助它们进食。须板能捕捉到微小的海洋动物，如浮游生物和磷虾。鲸抬起舌头，把浮游生物和磷虾关在嘴中。

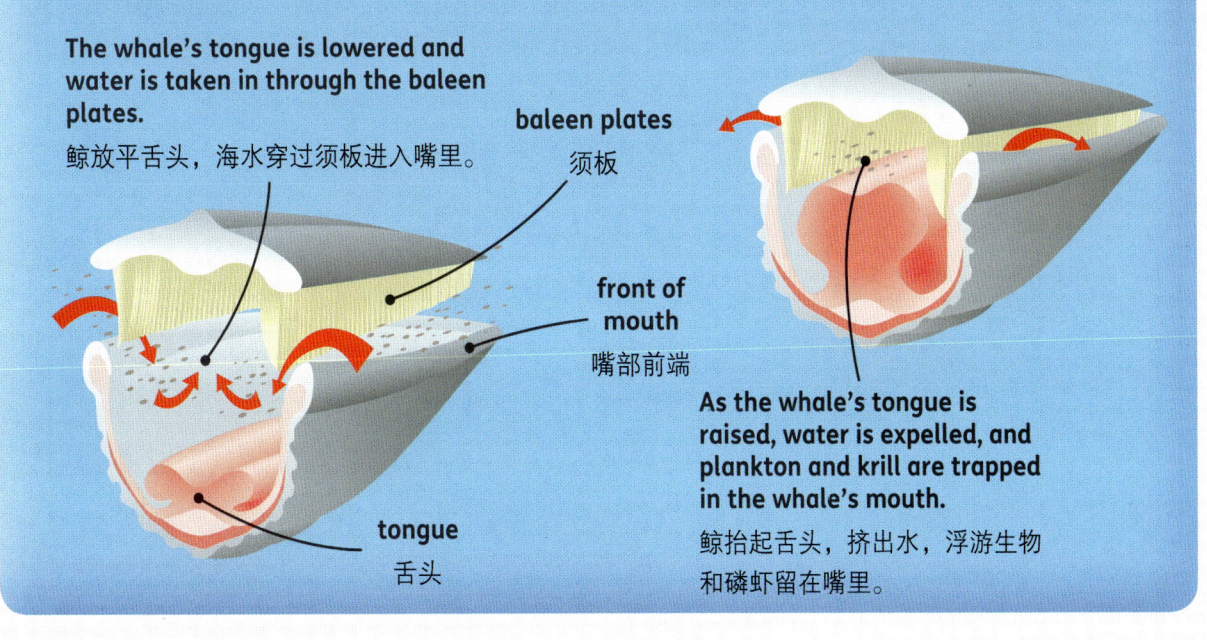

The whale's tongue is lowered and water is taken in through the baleen plates.
鲸放平舌头，海水穿过须板进入嘴里。

baleen plates
须板

front of mouth
嘴部前端

tongue
舌头

As the whale's tongue is raised, water is expelled, and plankton and krill are trapped in the whale's mouth.
鲸抬起舌头，挤出水，浮游生物和磷虾留在嘴里。

More ocean life 另一些海洋动物

Life in the ocean 海洋动物

Coral reefs 珊瑚礁

Coral reefs are made from living creatures! Coral polyps grow a hard outer case. When the coral dies, the casing remains and gradually builds up to form a reef. Coral reefs are home to millions of species of sea creatures. Many of these creatures have brilliant colours to blend in with the brightly coloured corals.

珊瑚礁是由动物形成的！珊瑚虫长出坚硬的外骨骼，死后骨架留了下来，久而久之形成一片珊瑚礁。珊瑚礁是各种海洋动物的家园。那儿有许多颜色鲜艳的动物，它们能混入五彩缤纷的珊瑚丛。

angelfish
神仙鱼

cowry
宝贝
The cowry is a species of sea snail.
宝贝是一种海洋贝类。

seahorse
海马
The male seahorse carries its young in a pouch.
雄性海马将小海马装在育儿袋里。

parrotfish
鹦嘴鱼

puffer fish
刺鲀（tún）
When it is threatened, the puffer fish expands by swallowing water.
受到威胁时，刺鲀吸水膨胀。

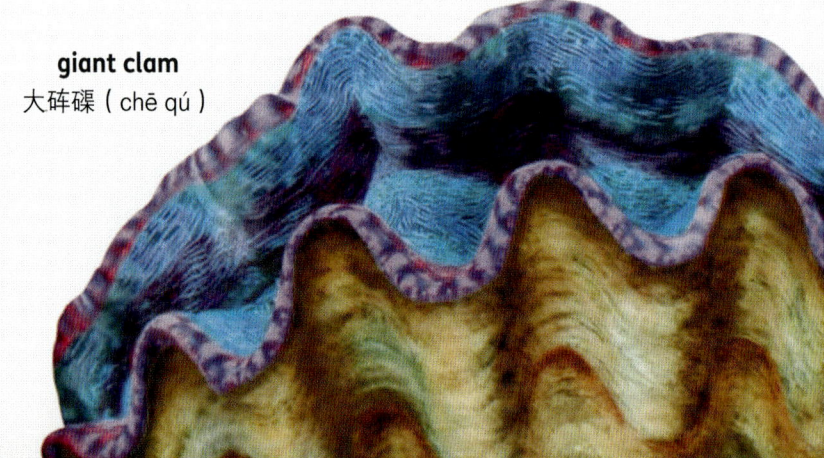

giant clam
大砗磲（chē qú）

Life in the ocean 海洋动物

green turtle
绿海龟

clown fish
小丑鱼

butterfly fish
蝴蝶鱼

triggerfish
鳞鲀

damselfish
雀鲷（diāo）

sea pen
海笔
Sea pens feed on plankton.
海笔以浮游生物为食。

brittle star
海蛇尾

sea goldie
丝鳍拟花鮨（yì）

lion fish
蓑鲉（yóu）

Coral reefs 珊瑚礁

93

Life in the ocean 海洋动物

Creatures of the deep 深海动物

It is almost completely dark on the ocean bed, but some deep-sea creatures can create their own light. Others have very large eyes to help them see. Sea sponges and tube worms stand on the sea bed. They are very simple creatures that look like plants.

海床几乎是完全黑暗的，但是有一些深海动物自己可以发光。其他动物则通过巨大的眼睛观察周围。海绵和管虫附着在海床之上。它们是最低等的动物，看上去像植物一样。

❶ hatchetfish 褶胸鱼

❷ vent fish 暖绵鳚

❸ lantern fish 灯笼鱼

❹ angler fish 深海鮟鱇

❺ giant tube worm 巨管虫

❻ deep-sea sponge 深海海绵

Life in the ocean 海洋动物

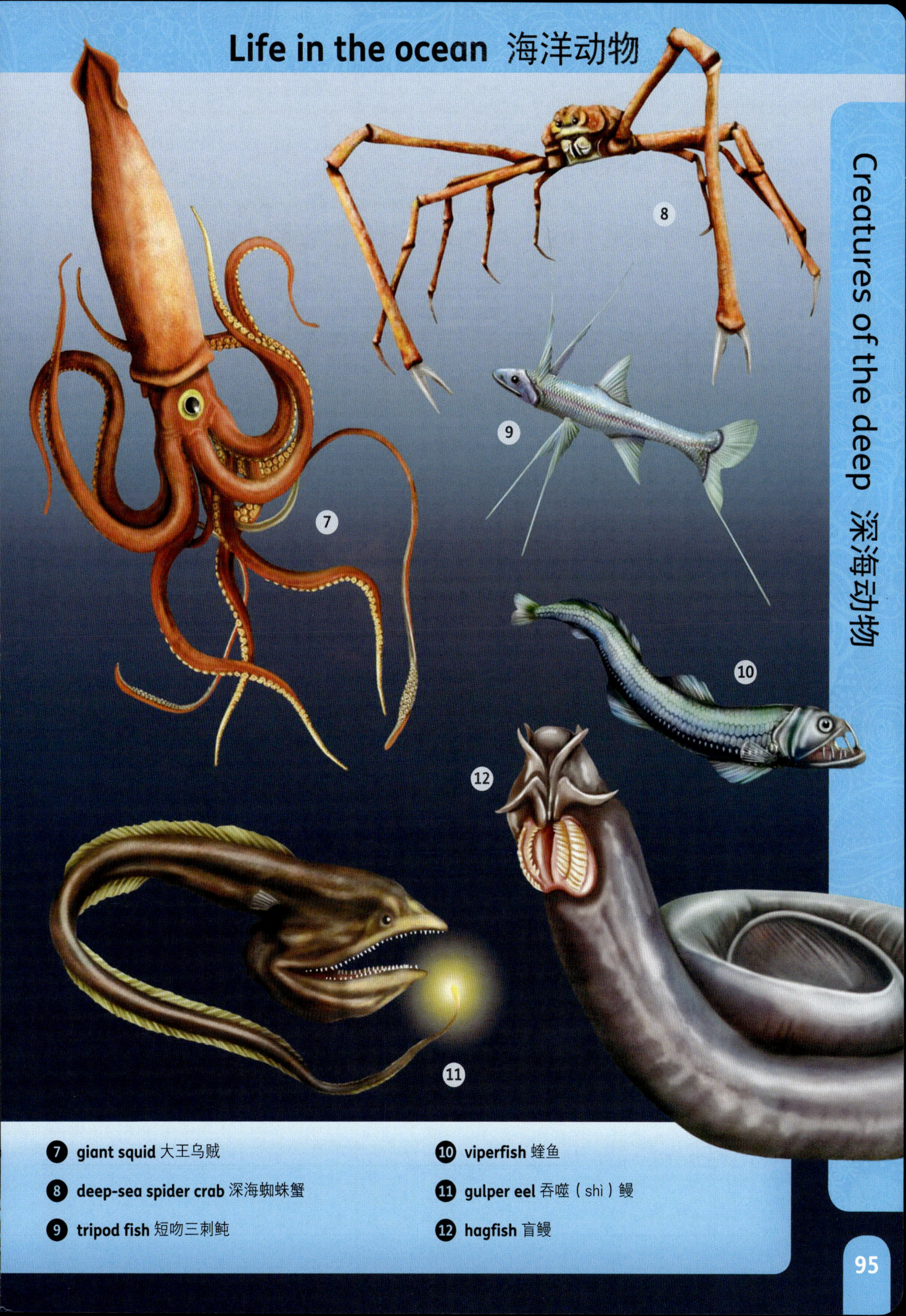

Creatures of the deep 深海动物

⑦ **giant squid** 大王乌贼
⑧ **deep-sea spider crab** 深海蜘蛛蟹
⑨ **tripod fish** 短吻三刺鲀
⑩ **viperfish** 蝰鱼
⑪ **gulper eel** 吞噬（shì）鳗
⑫ **hagfish** 盲鳗

95

On the Arctic ice cap, polar bears go hunting for seals. Gulls circle overhead and whales, seals and walruses swim in the icy seas...

在北极冰盖上,北极熊捕食海豹。鸥在空中盘旋,鲸、海豹、海象在冰冷的海水里游曳……

ANIMALS OF THE POLAR REGIONS
极地动物

Animals of the polar regions 极地动物

Polar habitats 极地栖息地

The Arctic and Antarctic regions are mostly ice and rock, and only a few species of animals can survive there. They are specially adapted to the low temperatures. At the edge of the ice are freezing seas and semi-frozen land, known as tundra.

北极和南极地区主要由冰层和岩石构成，只有少数动物能在那儿生存。它们适应了低温环境。冰层的边缘是寒冷的大海和半冻结的陆地，也就是苔原。

The Arctic 北极地区

In the Arctic, polar bears live on the ice and hunt seals.

在北极地区，北极熊生活在冰层上，捕食海豹。

polar bear 北极熊

beluga whale 白鲸

ivory gull 白鸥

Antarctica 南极洲

In Antarctica, penguins raise their young on the ice.

在南极洲，企鹅在冰层上抚育幼鸟。

emperor penguin 帝企鹅

Weddell seal 韦德尔海豹

albatross 信天翁

98

Animals of the polar regions 极地动物

The polar regions 极地分布图

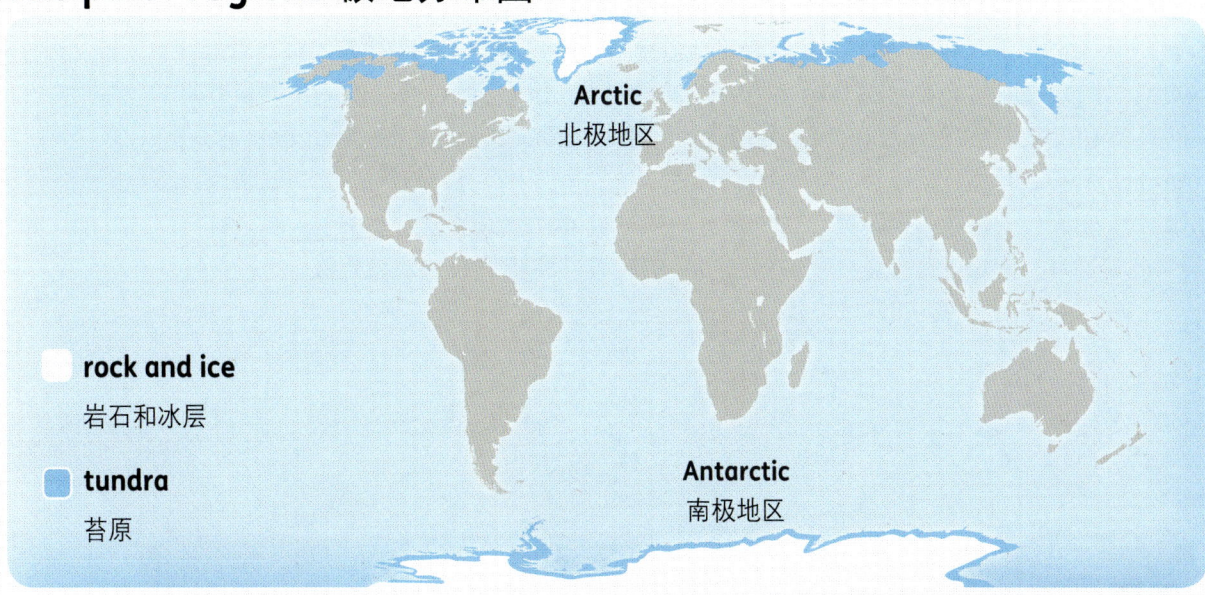

Arctic
北极地区

rock and ice
岩石和冰层

tundra
苔原

Antarctic
南极地区

Tundra regions 苔原地区

The tundra regions are covered in snow for the most of the year.
苔原地区在一年的大部分时间里都有积雪。

Arctic wolf
北极狼

Canada goose
加拿大雁

ermine
白鼬

Polar habitats 极地栖息地

99

Animals of the polar regions 极地动物

Arctic and Antarctic wildlife 北极和南极的动物

Seals are found in the Arctic and in Antarctica. Whales, fish and other sea creatures swim in the icy seas and birds fly overhead looking for food. Several species of penguins live in Antarctica but no penguins are found in the Arctic regions. The polar bear is only found in the Arctic.

北极地区和南极洲都有海豹。鲸、鱼类和其他海洋动物在冰冷的海水里游曳，鸟儿在空中盘旋觅食。南极洲有多种企鹅，但北极地区没有。北极熊只生活在北极。

Arctic skua
北极贼鸥

walrus
海象

ringed seal
环斑海豹

Seals are protected from the cold by a thick layer of fat known as blubber.
海豹长了一身厚厚的脂肪抵御寒冷，这种脂肪叫海兽脂。

rockhopper penguin
凤头黄眉企鹅

Arctic skate
北极钝头鳐

Animals of the polar regions 极地动物

harp seal
竖琴海豹

snow petrel
雪鹱（hù）

narwhal
独角鲸
The narwhal's tusk is really a long tooth.
独角鲸的角其实是一根长牙。

sperm whale
抹香鲸

king penguin
王企鹅
Male penguins keep their eggs warm.
雄性企鹅孵蛋。

elephant seal
象海豹
Male elephant seals are ferocious fighters.
雄象海豹打架非常凶狠。

Arctic and Antarctic wildlife 北极和南极的动物

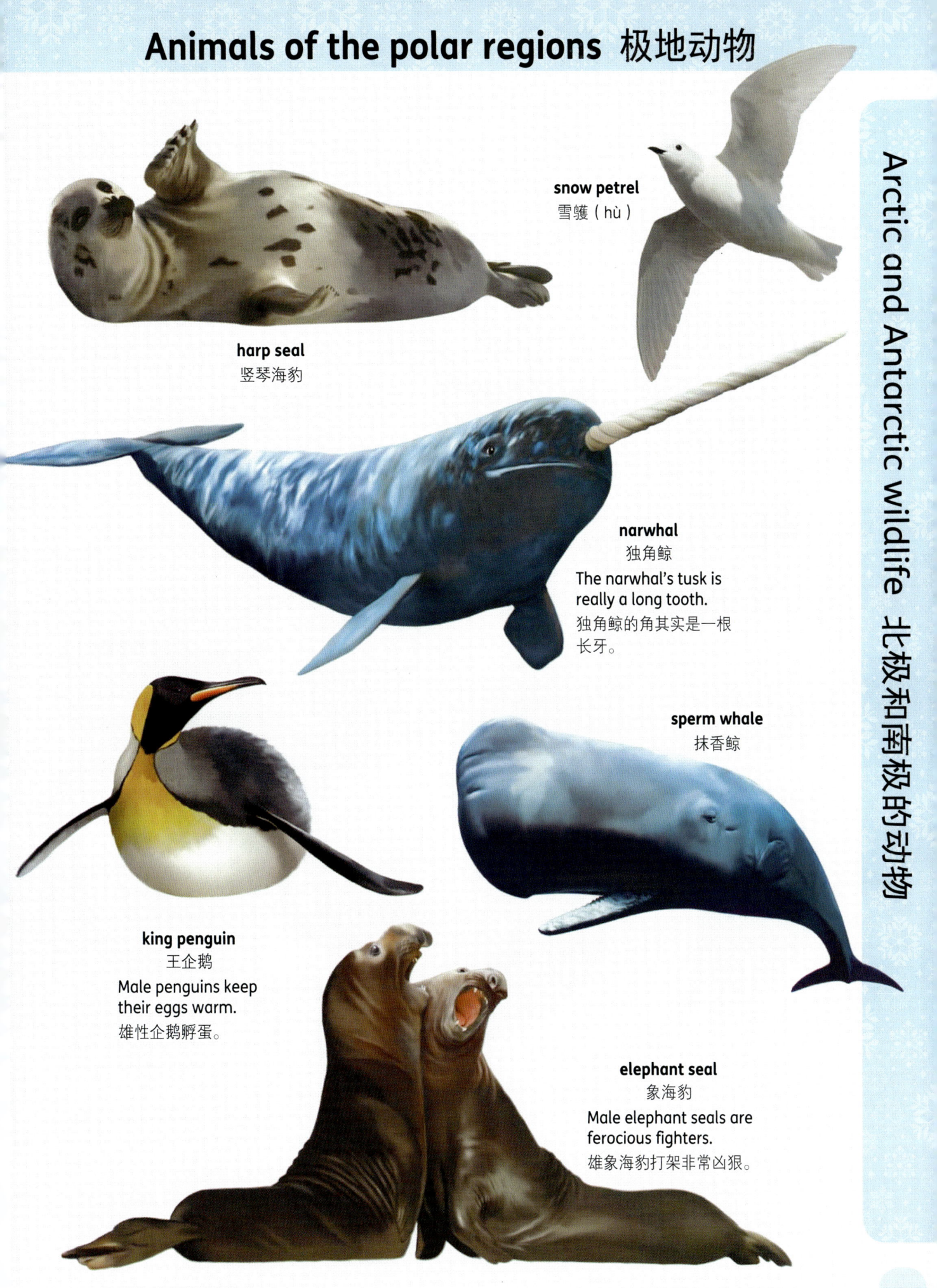

Animals of the polar regions 极地动物

Life in the tundra 苔原动物

Tundra regions have very short summers. In these warmer months, the snow disappears and the icy ground becomes soft. Some animals shed their winter coats and some emerge from burrows where they have sheltered.

苔原地区夏季十分短暂。在温暖的月份，积雪融化，冰冻的地面变得松软。有的动物褪去冬季的皮毛，有的钻出它们藏身的洞穴。

Arctic hare
北极兔

caribou
驯鹿
Caribou are also known as reindeer.
驯鹿还有个名字叫 reindeer。

snow goose
雪雁

Arctic lemming
北极旅鼠

eider duck
绒鸭
Eider ducks have very soft feathers called down that they use to line their nests.
绒鸭有非常柔软的羽毛，叫绒羽，它们把绒羽铺在窝里。

Animals of the polar regions 极地动物

grizzly bear 灰熊

snowy owl 雪鸮

Adaptations to snow 适应冰雪

Arctic fox 北极狐

summer 夏季

extra body fat 脂肪增多

white coat for camouflage 白色皮毛起伪装作用

winter 冬季

fur on base of paws 脚底长出软毛

musk ox 麝牛

summer 夏季

long, shaggy overcoat 又厚又长的上层皮毛

winter 冬季

fleecy undercoat 蓬松的下层皮毛

heavy hooves to break through snow 宽大的蹄子能在雪中行走

Life in the tundra 苔原动物

103

Widespread creatures 分布广泛的动物

Some animals are found in a wide range of habitats across the world. These widespread creatures include many species of insects and minibeasts. Insects often live on food or plants, and this means they are easily transported around the world.

有的动物在世界各地的多种栖息地可以看到。这些分布广泛的动物包括多种昆虫和其他无脊椎动物。昆虫一般以多种食物或植物为食，因此很容易传播到世界各地。

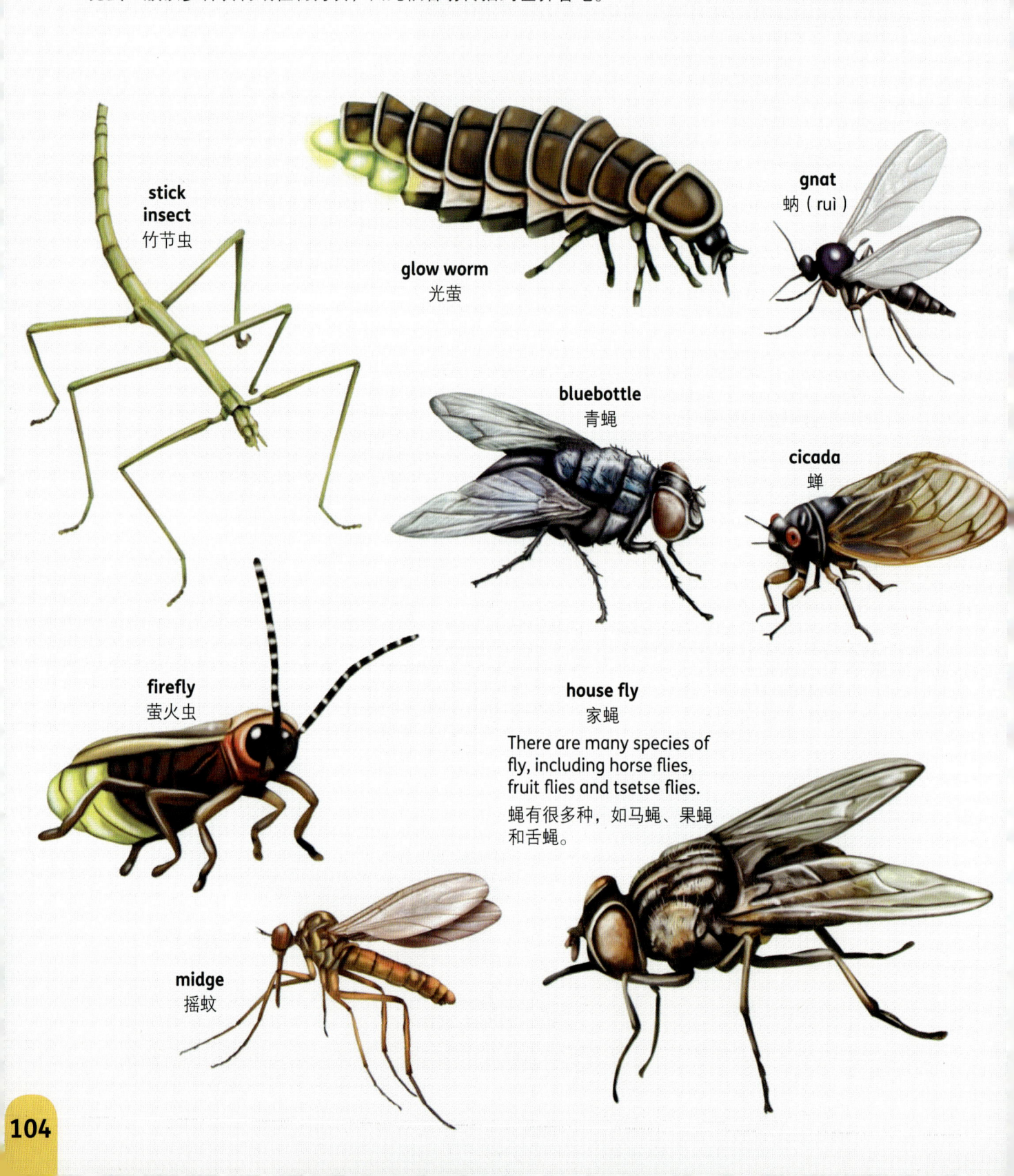

stick insect 竹节虫

glow worm 光萤

gnat 蚋（ruì）

bluebottle 青蝇

cicada 蝉

firefly 萤火虫

house fly 家蝇

There are many species of fly, including horse flies, fruit flies and tsetse flies.
蝇有很多种，如马蝇、果蝇和舌蝇。

midge 摇蚊

Widespread creatures 分布广泛的动物

crane fly
大蚊
Crane flies are often known as daddy longlegs.
大蚊又被称作长腿爸爸。

cabbage white butterfly
菜粉蝶

cricket
蟋蟀

wasp
胡蜂

hornet
大胡蜂

ladybird
瓢虫

earwig
蠼螋（qú sōu）

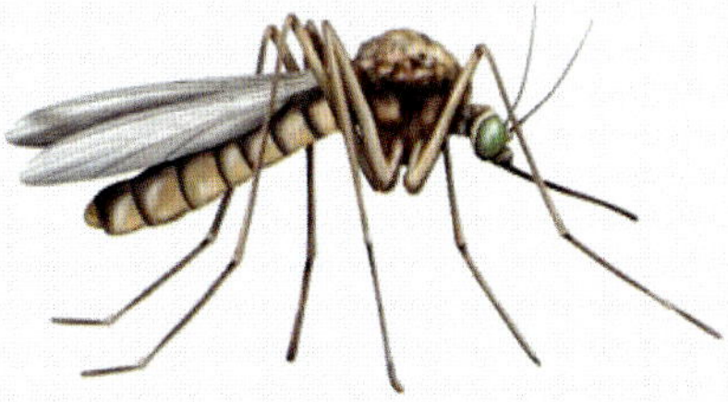

mosquito
蚊子

cockroach
蟑（zhāng）螂

dung beetle
蜣（qiāng）螂

105

Widespread birds 分布广泛的鸟类

Some types of bird are found in many parts of the world. Here are some common examples. (Birds in different countries sometimes have the same name but do not belong to the same species. The robin and the blackbird are examples of this.)

有些鸟类在世界很多地方可以看到，这两页中是一些常见的例子。（有时不同国家的鸟儿名字相同，却并不属于同一个物种。如欧亚鸲（qú）和旅鸫（dōng）都叫 robin，乌鸫和拟鹂都叫 blackbird。）

jackdaw
寒鸦
Jackdaws belong to the crow family.
寒鸦属于鸦科。

chaffinch
苍头燕雀
The chaffinch belongs to the finch family. Other finches include the bullfinch and the goldfinch.
苍头燕雀属于雀科，雀科的其他鸟儿还有灰雀和金翅雀。

cuckoo
杜鹃

thrush
鸫

peacock
孔雀
Peacocks are domesticated in many parts of the world. They live in the wild in Africa and Asia.
孔雀在世界很多地方都已经驯化，非洲和亚洲还有野生孔雀。

Widespread birds 分布广泛的鸟类

skylark
云雀

sparrow
麻雀

starling
椋（liáng）鸟

common blackbird/ Eurasian blackbird
乌鸫

swan
天鹅
Australian swans are black.
澳大利亚的天鹅是黑色的。

wren
鹪鹩（jiāo liáo）

pigeon
鸽子
Pigeons are found in large numbers in cities.
城市里有大量鸽子。

collared dove
斑鸠（jiū）

magpie
喜鹊

107

Vocabulary builder 扩展词汇

What do you call a baby animal?
如何称呼幼年的动物？

ANIMAL 动物	YOUNG 幼崽	ANIMAL 动物	YOUNG 幼崽
antelope 羚羊	calf	eagle 雕	eaglet
badger 獾	cub	eel 鳗	elver
bear 熊	cub	elephant 象	calf
beaver 河狸	kit	ferret 雪貂	kit
bobcat 短尾猫	kitten	fish 鱼	fry
buffalo 水牛	calf	frog 蛙	tadpole
camel 骆驼	calf	giraffe 长颈鹿	calf
caribou 驯鹿	fawn	goose 鹅；雁	gosling
cougar 美洲狮	kitten	hare 野兔	leveret
coyote 郊狼	puppy	horse 马	foal / colt (male) / filly (female)
dog 狗	puppy / pup / whelp		

Vocabulary builder 扩展词汇

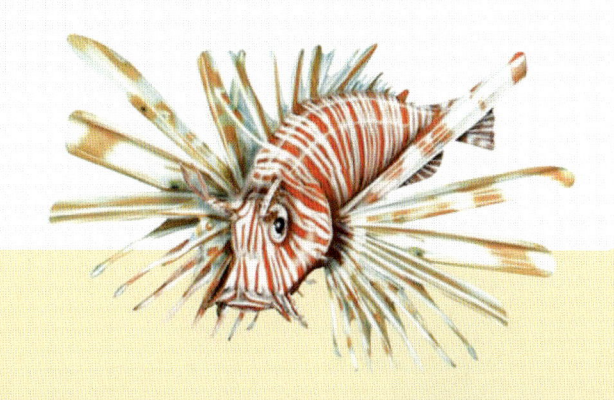

ANIMAL 动物	YOUNG 幼崽
kangaroo 袋鼠	*joey*
lion 狮	*cub*
owl 猫头鹰，鸮	*owlet*
pigeon 鸽子	*squab / squeaker*
pike 狗鱼	*pickerel*
rhinoceros 犀牛	*calf*
salmon 鲑	*parr / smolt*
seal 海豹	*calf / pup*
spider 蜘蛛	*spiderling*

ANIMAL 动物	YOUNG 幼崽
swan 天鹅	*cygnet*
tiger 虎	*cub*
toad 蟾蜍	*tadpole*
wallaby 沙袋鼠	*joey*
walrus 海象	*cub*
weasel 伶鼬	*kit*
whale 鲸	*calf*
wolf 狼	*cub / pup / whelp*
zebra 斑马	*foal*

Vocabulary builder 扩展词汇

What do you call a group of...?
如何称呼一群动物？

ants 蚂蚁	*a colony*
bees 蜜蜂	*a swarm*
birds 鸟	*a flock*
dolphins 海豚	*a school*
fish 鱼	*a shoal / a school*
geese 鹅；雁	*a gaggle / a flock*
lions 狮子	*a pride*
monkeys 猴子	*a troop*
whales 鲸	*a pod*
wolves 狼	*a pack*

Vocabulary builder 扩展词汇

What noise does an animal make?
动物发出什么样的叫声？

Bees buzz. 蜜蜂嗡嗡作响。

Birds sing, tweet, warble and chirp. 鸟儿啼啭、叽叽、喳喳、啁（zhōu）啾。

Cows moo. 奶牛哞哞叫。

Dogs bark and growl. 狗汪汪吠叫或低吼。

Donkeys bray. 驴嘶叫。

Ducks quack. 鸭子嘎嘎叫。

Elephants trumpet. 大象嗷嗷吼叫。

Frogs croak. 青蛙呱呱叫。

Hens cluck. 母鸡咯咯叫。

Horses neigh and whinny. 马儿咴（huī）咴嘶鸣。

Lions roar. 狮子咆哮。

Mice squeak. 老鼠吱吱叫。

Monkeys chatter. 猴子吱吱叫。

Owls hoot. 猫头鹰呜呜叫。

Parrots screech. 鹦鹉喳喳叫。

Snakes hiss. 蛇发出嘶嘶声。

Wolves howl. 狼长嚎。

111

Vocabulary builder 扩展词汇

Animal word origins 动物词语的词源

Some animal names have interesting origins. They come from different languages, perhaps from the way the animal looks, like "rhinoceros", or the noise that it makes, such as "cuckoo". The technical name for the study of word origins is "etymology". Here are the origins of some of the names of animals you can find in this book.

有些动物名称的来源十分有趣。它们来自不同的语言，有的动物是因为长相而得名，如犀牛，有的因为叫声得名，如杜鹃。研究词语来源的学科叫"词源学"。下面列举了一些本书中出现的动物名称的来源。

aardvark is from Afrikaans, from *aarde* meaning "earth" and *vark* meaning "pig"
aardvark（土豚）源自阿非利堪斯语，由 aarde（土）和 vark（猪）两个词组合而成

alligator is from Spanish, from *el lagarto* meaning "the lizard"
alligator（短吻鳄）源自西班牙语的 el lagarto（蜥蜴）

anemone is from a Greek word meaning "windflower" (from the belief that the flower opens when it is windy)
anemone（海葵）源自希腊语中表示"风花"的词（人们以为这种花在刮风时开放）

animal is from Latin *animalis* meaning "having breath"
animal（动物）源自拉丁语的 animalis（有呼吸）

antelope is from a late Greek word *antholops* (which was originally the name of a mythical creature)
antelope（羚羊）源自希腊语的 antholops（最初指一种神话中的动物）

armadillo is from Spanish *armado* meaning "armed man"
armadillo（犰狳）源自西班牙语的 armado（拿武器的人）

badger is perhaps from "badge", because of the markings on a badger's head. The word dates from the 16th century; the earlier Old English word for a badger was "brock".
badger（獾）可能源自 badge（标记）一词，因为獾的头上有黑白相间的条纹。badger 最早可追溯到 16 世纪；在那之前，獾在古英语中称为 brock。

barnacle first referred to a kind of goose which was once believed to hatch from shellfish attached to rocks
barnacle（藤壶）最初指一种雁，人们曾经以为那种雁是从岩石上的贝壳里孵出来的

basilisk is from Greek *basilikos* which originally meant "little king" and is related to our words "basil" and "basilica"
basilisk（冠蜥）源自希腊语 basilikos 一词，最初指"小国王"，也与英语中的 basil（罗勒）和 basilica（会堂）有联系

112

Vocabulary builder 扩展词汇

beaver is from an Old English word *beetle* which is from Old English *bitula*, from bitan "to bite", because of its biting mouthparts
beaver（河狸）源自古英语的 beetle 一词，而 beetle 源自古英语 bitula，bitula 又源自 bitan（咬），河狸因其吻部而得名

budgerigar is from Australian Aboriginal *budgeri* meaning "good" and *gar* meaning "cockatoo"
budgerigar（虎皮鹦鹉）由澳大利亚土著语的 budgeri（好的）和 gar（凤头鹦鹉）组合而成

buzzard is from Latin *buteo* meaning "falcon"
buzzard（鵟）源自拉丁语的 buteo（鵟）

caterpillar is from Old French *chatepelose* meaning "hairy cat"
caterpillar（毛虫）源自古法语的 chatepelose（毛茸茸的猫）

centipede is from *centi-* and a Latin word *pedes* meaning "feet"
centipede（蜈蚣）是 centi（百）和拉丁语词 pedes（足）的组合

crocodile is from a Greek word *krokodilos* meaning "worm of the stones"
crocodile（鳄鱼）源自希腊语的 krokodilos（石头虫）

cuckoo is named after the sound of its call
cuckoo（杜鹃，亦称布谷鸟）因它的叫声而得名

dromedary camel is from Greek *dromas* meaning "runner"
dromedary camel（单峰驼）源自希腊语的 dromas（奔跑者）

ferret is from Latin *fur* meaning "thief"
ferret（雪貂）源自拉丁语的 fur（小偷）

flamingo is from a Spanish word *flamengo*, and is connected to the Latin *flamma* meaning "a flame" because of its colour
flamingo（红鹳）源自西班牙语的 flamengo，与拉丁语的 flamma（火焰）有关，这是因为它们火红的颜色

gorilla is from a Greek word, which is probably from an African word denoting a wild or hairy person
gorilla（大猩猩）源自希腊语单词，可能最早源自非洲语言中表示"野人"或"多毛的人"的词

halibut is from "holy" and *butt*, a dialect word meaning "flatfish" (because it was eaten on Christian holy days, when meat was forbidden)
halibut（大比目鱼）由 holy（神圣）和 butt（"比目鱼"的方言说法）组合而成（因为过去在基督教节日时不允许吃肉，人们常吃这种鱼）

hippopotamus is from Greek *hippos* meaning "horse" and *potamos* meaning "river"
hippopotamus（河马）由希腊语的 hippos（马）和 potamos（河）组合而成

kangaroo is an Australian Aboriginal word
kangaroo（袋鼠）是澳大利亚土著语中的单词

113

Vocabulary builder 扩展词汇

lobster is via Old English from a Latin word *locusta* meaning "crustacean" or "locust"
lobster（龙虾）是古英语从拉丁语中借用来的词，源自拉丁语中的 locusta（甲壳动物；蝗虫）

penguin is possibly from Welsh *pen gwyn* meaning "white head"
penguin（企鹅）可能源自威尔士语的 pen gwyn（白头）

piranha comes via Portuguese from Tupi (a South American language)
piranha（食人鲳）是葡萄牙语从图皮语（一种南美洲的语言）中借用来的词

plaice is from Greek *platys* meaning "broad"
plaice（鲽）源自希腊语的 platys（宽的）

porcupine is from Old French *porc espin* meaning "spiny pig"
porcupine（豪猪）源自古法语的 porc espin（带刺的猪）

porpoise is from the Latin words *porcus* meaning "pig" and *piscis* meaning "python" which is from Python, the name of a huge serpent killed by Apollo in Greek legend
porpoise（鼠海豚）由拉丁语的 porcus（猪）和 piscis（蟒）组合而成，而 piscis 又源自 Python（皮同），是希腊神话中被太阳神阿波罗杀死的巨蟒

reptile is from Latin *reptilis* meaning "crawling"
reptile（爬行动物）源自拉丁语的 reptilis（爬行）

rhinoceros is from Greek *rhinos* meaning "of the nose" and *keras* meaning "horn"
rhinoceros（犀牛）由希腊语的 rhinos（鼻子的）和 keras（角）组合而成

snake is from an Old English word *snaca* meaning "to crawl" or "to creep"
snake（蛇）源自古英语的 snaca（爬行）

spider is from an Old English word *spithra* meaning "spinner"
spider（蜘蛛）源自古英语的 spithra（纺织者）

squirrel is from Greek *skiouros* which is from *skia* meaning "shadow" and *oura* meaning "tail" (because its long bushy tail cast a shadow over its body and kept it cool)
squirrel（松鼠）源自希腊语的 skiouros，由 skia（影子）和 oura（尾巴）组合而成（因为松鼠蓬松的长尾巴能遮阳，帮身体降温）

tadpole is from toad and an old word *poll* meaning "head"
tadpole（蝌蚪）由 toad（蟾蜍）和古词 poll（头部）组合而成

tarantula is from Taranto in southern Italy, because the spider's bite was thought to cause tarantism, a psychological illness once common in southern Italy
tarantula（狼蛛）源自意大利南部城市名 Taranto（塔兰托），因为人们曾经认为被这种蜘蛛咬后会得毒蛛病，一种意大利南部常见的精神病

wildebeest is an Afrikaans word meaning "wild beast"
wildebeest（角马）是阿非利堪斯语单词，意思是 wild beast（野兽）

Vocabulary builder 扩展词汇

Animal idioms 动物习语

An idiom is a phrase or group of words that have a special meaning that is not obvious from the words themselves, for example, "to be in hot water" means to be in trouble or difficulty. Here are some idioms which feature some of the animals in this book.

习语是指搭配使用、有特殊意义的短语或词组，如 be in hot water（在热水中）的意思是"惹上麻烦"或"遇到困难"。下面是一些与本书中出现过的动物有关的习语。

a busy bee
a very busy person
字面义 一只忙碌的蜜蜂
比喻义 大忙人

keep the wolf from the door
to avoid going hungry
字面义 拒狼于门外
比喻义 勉强度日；糊口

on a wild goose chase
To be on a wild goose chase is to be searching for something that is impossible to find, so that you waste a lot of time.
字面义 追赶大雁
比喻义 "追赶大雁"是指徒劳的寻找，白费时间的追寻。

like a fish out of water
If you are like a fish out of water, you feel awkward because you are in a situation that is not at all familiar.
字面义 像离开了水的鱼
比喻义 "像离开了水的鱼"指一个人来到陌生环境，感到不得其所。

have a bee in your bonnet
If you have a bee in your bonnet about something, you cannot stop thinking or talking about it because you think it is very important.
字面义 帽子里有只蜜蜂
比喻义 "帽子里有只蜜蜂"是说你对某件事念念不忘，你没法不去想它，因为这件事在你心中非常重要。

it's the bee's knees
If you think something is really excellent, you can say "it's the bee's knees".
字面义 是蜜蜂的膝盖
比喻义 出类拔萃的人（或物）。如果你想说某件事非常棒，可以说 it's the bee's knees。

hold your horses
wait a moment, don't be so hasty
字面义 控制住你的马儿
比喻义 且慢；别仓促行动

a leopard can't change its spots
people can't change their basic nature
字面义 豹子没法改变自己的斑纹
比喻义 本性难移

put the cart before the horse
to do things in the wrong order
字面义 把车放在马前面
比喻义 本末倒置

straight from the horse's mouth
Information that is straight from the horse's mouth is from the person who is most directly involved and so is likely to be accurate. It is as if a horse in a horse race was telling you which horse was going to win the race.
字面义 直接来自马儿的嘴里
比喻义 一条消息直接来自马儿的嘴里，是指消息由直接参与者提供，因而十分可靠。就像参加赛马的马儿告诉你哪匹马儿将赢得比赛。

the straw that broke the camel's back
a small difficulty that, coming on top of a lot of other difficulties, makes a situation too much to bear
字面义 压垮骆驼的最后一根稻草
比喻义 在许多困难之上再加上一个小困难，情况因而变得令人难以承受

115

Animal detective quiz 动物知识小测验

How much do you know about animals? 你对动物了解多少？

Try this animal detective quiz and track down the creatures in the pages of this book. The answers are at the back of the book.

做一做下面的测试题，在书中寻找相应的动物。正确答案见书后。

Animal hunters 动物捕食者

1. Which creature injects its victims with deadly saliva?

   哪种动物会向它的猎物注射致命的毒液？

2. Which creature sucks blood from larger animals?

   哪种动物会吸食大型动物的血液？

3. Which creature stuns its prey with an electric shock ?

   哪种动物会用电流击晕它的猎物？

Watch out! 当心！

4. Which animal warns off attackers by producing a very strong smell?

   哪种动物使用强烈的臭味警告敌人？

5. Which animal warns off attackers by swallowing water to make itself look larger?

   哪种动物会吞水使自己变大以警告敌人？

6. Which animal warns off attackers by displaying warning colours to show it is poisonous?

   哪种动物展示警戒色以警告敌人它是有毒的？

Special features 特征

7. Which creature has bony plates covering its body?

   哪种动物有骨板覆盖身体？

8. Which creature has enormous ears that can lose heat fast?

   哪种动物有巨大的耳朵以便快速散热？

9. Which creature has a long tongue for sucking nectar?

   哪种动物有长舌可以吸食花蜜？

Animal detective quiz 动物知识小测验

Unusual behaviour 不寻常的行为

10. Which animal can run on the surface of water?
哪种动物能在水面上飞奔？

11. Which animal communicates by using clicks and whistles?
哪种动物用吱吱声和口哨声相互交流？

12. Which animal sleeps for around 20 hours a day?
哪种动物一天能睡 20 个小时左右？

Animal homes 动物的家

13. Which creature lives in a lodge?
哪种动物住在自己用树枝在水上修筑的巢穴里？

14. Which creature builds a home from earth and saliva?
哪种动物用泥土和唾液筑巢？

15. Which creature lives inside the guts of other animals?
哪种动物住在其他动物的肠道中？

Animal parts 动物的身体部分

16. Which animal has a spinneret?
哪种动物有丝囊？

17. Which animal has a dewlap?
哪种动物有肉垂？

18. Which animal has a siphon?
哪种动物有虹吸管？

What am I? 我是哪种动物？

19. I am a mammal that lays eggs.
我是产卵的哺乳动物。

20. I am a fish that can breathe on land.
我是能在陆地上呼吸的鱼。

English-Chinese index 英汉索引

aardvark 土豚　58
abdomen 腹部　22, 24
adder 蝰　55
African elephant 非洲象　57
African forest elephant
非洲森林象　31
albatross 信天翁　98
alligator 短吻鳄　74
alpaca 小羊驼　51
Amazon parrot 亚马孙鹦哥　34
Amazon rainforest 亚马孙雨林
34-35
American bullfrog 美洲牛蛙　72
amphibian 两栖动物　20
anaconda 水蚺　68
anemone 海葵　86
angelfish 神仙鱼　92
angler fish 深海鮟鱇　94
ant 蚂蚁　31, 80
Antarctic 南极地区　98
Antarctica 南极洲　98
anteater 食蚁兽　28
antelope 羚羊　57
antenna 触角　22, 24
ape 类人猿　15, 29, 31, 32, 49
Arctic 北极地区　98-101
Arctic fox 北极狐　103
Arctic hare 北极兔　102
Arctic lemming 北极旅鼠　102
Arctic wolf 北极狼　99
arm 臂；腕足　15, 23
armadillo 犰狳　59
army ant 行军蚁　30
assassin bug 猎蝽　30
Australian bush 澳大利亚荒野
55, 62-63
baby 幼崽　15
back 背部　14
Bactrian camel 双峰驼　79

badger 獾　42
baleen 鲸须　91
bandicoot 袋狸　63
barnacle 藤壶　86
barracuda 大舒　88
basilisk lizard 双嵴冠蜥　73
beak 喙；嘴　16, 23
bear 熊　44, 98, 103
bearded vulture 胡兀鹫　48
beaver 河狸　69
beaver lodge 河狸巢穴　69
bee 蜜蜂　25
beehive 蜂窝　25
beetle 甲虫　24, 29, 30, 33
bell 伞状体　23
beluga whale 白鲸　98
Bengal tiger 孟加拉虎　75
big cat 大型猫科动物　28, 30, 39, 44,
50, 56, 57, 75
bilby 兔耳袋狸　63
bill 喙；鸟嘴　17
bird 鸟　16-17
bison 野牛　54
black bear 黑熊　44
black widow spider 黑寡妇蜘蛛　81
bladder 膀胱　15
blue tang 黄尾副刺尾鱼　84
blue whale 蓝鲸　91
blue-tongued skink 蓝舌蜥　62
bobcat 短尾猫　39
brain 脑　15
breast 胸部　17
brimstone butterfly 硫磺蝶　38
budgerigar 虎皮鹦鹉　63
buffalo 水牛　56
bug 虫　30
bullfrog 牛蛙　72
Burmese python 缅甸蟒　19
bush 灌木；荒野　62-63

butterfly 蝴蝶　25, 28, 38
butterfly fish 蝴蝶鱼　93
buzzard 鵟　50
caiman 凯门鳄　67
camel 骆驼　79, 80
camouflage 保护色　19
Canada goose 加拿大雁　99
canopy 树冠层　28, 32-33
capuchin monkey 卷尾猴　32
capybara 水豚　72
carapace 甲壳　22
caribou 驯鹿　102
carp 鲤鱼　73
cassowary 鹤鸵　31
caterpillar 毛虫　25
catfish 鲇鱼　66
caudal fin 尾鳍　22
cavy 豚鼠　51
centipede 蜈蚣　42
chamois 臆羚　50
cheetah 猎豹　57
chest 胸部　15
chimpanzee 黑猩猩　29
chinchilla 毛丝鼠　49
Chinese water dragon lizard
长鬣蜥　19
chipmunk 花鼠　38
chrysalis 蛹　25
claw 爪；螯　14, 18, 22
clouded leopard 云豹　28
clown fish 小丑鱼　93
coatimundi 长鼻浣熊　34
cockatoo 凤头鹦鹉　63
cod 鳕　89
cold desert 冷荒漠　79
colobus monkey 疣猴　33
colon 结肠　15
colugo 鼯猴　33
common adder 极北蝰　55

English-Chinese index 英汉索引

common buzzard 普通鵟 50
common carp 普通鲤鱼 73
common frog 黑斑蛙 70
common toad 普通蟾蜍 72
coot 骨顶鸡 66
coral reef 珊瑚礁 84, 92-93
coral snake 珊瑚蛇 29
cormorant 鸬鹚 86
cougar 美洲狮 50
cowry 宝贝 92
coyote 郊狼 60
coypu 河狸鼠 69
crab 螃蟹 86, 95
crane 鹤 74
crayfish 淡水螯虾 68
crimson rosella 红玫瑰鹦鹉 49
crocodile 鳄鱼 18, 68
crusher claw 破碎螯 22
cuttlefish 乌贼 84
damselfish 雀鲷 93
damselfly 豆娘 66
deciduous forest 落叶林 39-41
deep-sea creatures 深海动物 94
deep-sea spider crab 深海蜘蛛蟹 95
deep-sea sponge 深海海绵 94
deer 鹿 38, 40
desert tortoise 阿氏沙龟 79
Desert wildlife 荒漠动物 76-81
dewlap 肉垂 20
diamondback terrapin 钻纹龟 75
dingo 澳洲野犬 62
dipper 河乌 70
dog 狗 14
dogfish 狗鲨 88
dolphin 海豚 90
domesticated creatures 驯化的动物 51
dormouse 睡鼠 42

dorsal fin 背鳍 22
dragonfly 蜻蜓 70
dromedary camel 单峰驼 80
drone 雄蜂 25
duck 鸭 17, 102
dugong 儒艮 90
eagle 雕 16, 29, 50
ear 耳朵 14
eardrum 鼓膜 20
earthworm 蚯蚓 42
eel 鳗 69, 95
egg 卵；蛋 16, 25, 59, 69
eider duck 绒鸭 102
electric eel 电鳗 69
elephant 象 31, 57
emerald tree boa 绿树蚺 29
emergent layer 林上层 28
emperor moth 帝王蛾 60
emperor penguin 帝企鹅 98
emu 鸸鹋 63
ermine 白鼬 99
evergreen forest 常绿林 39
eye 眼 14, 23
eyelash viper 睫角棕榈蝮 32
falcon 隼 50
fallow deer 黇鹿 38
feather 羽毛 16
fennec fox 聊狐 78
ferret 雪貂 61
fin 鳍 22
finger 手指 15
fish and other sea creatures 鱼类和其他海洋动物 22-23, 82-95
flamingo 红鹳 72
flank 胁腹 14
flipper 鳍足；鳍肢 18
Florida panther 佛罗里达美洲狮 75
flounder 偏口鱼 88
flying dragon lizard 飞蜥 28

flying fish 飞鱼 85
flying gecko 褶虎 33
foot 足 15
foreleg 前腿 14
Forest and woodland wildlife 森林和林地动物 36-45
forest floor 林地层 28
forewing 前翅 25
forked tongue 叉状舌 19
fox 狐狸 41, 78, 103
frog 蛙 20, 21, 28, 33, 70, 72
frogspawn 蛙卵 21
funnel web spider 漏斗网蛛 30
galah 粉红凤头鹦鹉 62
garden snail 花园蜗牛 38
gazelle 瞪羚 54
gecko 壁虎 33
gerbil 沙鼠 80
giant anteater 大食蚁兽 28
giant clam 大砗磲 92
giant panda 大熊猫 51
giant rainforest praying mantis 澳洲斧螳 32
giant squid 大王乌贼 95
giant tube worm 巨管虫 94
gibbon 长臂猿 29
gill 鳃 21, 23
giraffe 长颈鹿 57
giraffe weevil 长颈象甲 33
goanna 澳洲巨蜥 55
goat 山羊 48
golden eagle 金雕 50
goldfish 金鱼 73
goliath beetle 巨花潜金龟 29
goliath spider 捕鸟蛛 31
goose 鹅；雁 17, 99
gorilla 大猩猩 31, 49
grasshopper 蚂蚱 61

119

English-Chinese index 英汉索引

Grassland wildlife 草原动物 52-63
great grey owl 乌林鸮 45
great hornbill 双角犀鸟 28
greater water boatman 松藻虫 70
green iguana 美洲鬣蜥 34
green turtle 绿海龟 93
grey squirrel 灰松鼠 41
grizzly bear 灰熊 103
grouse 松鸡 60
guinea fowl 珍珠鸡 60
gulper eel 吞噬鳗 95
haddock 黑线鳕 89
hagfish 盲鳗 95
halibut 大比目鱼 88
hammerhead shark 双髻鲨 90
hare 野兔 61, 102
harpy eagle 美洲角雕 29
hatchetfish 褶胸鱼 94
hawk owl 鹰鸮 39
head 头部 23, 24
heart 心脏 15
hedgehog 刺猬 42
Hercules beetle 巨犀金龟 30
Hermann's tortoise 赫氏陆龟 19
heron 鹭 17, 70
herring 鲱 85
hind leg 后腿 14
hindwing 后翅 25
hippopotamus 河马 66
hollow fang 中空的毒牙 19
honey bee 蜜蜂 25
honeyeater 吸蜜鸟 32
honeypot ant 蜜壶蚁 80
hood 皮褶 19
hoof 蹄 14
hornbill 犀鸟 28
horned desert viper 角蝰 81
horse 马 14
hot desert 热荒漠 78

howler monkey 吼猴 34
hummingbird 蜂鸟 33
humpback whale 座头鲸 91
ibex 羱羊 51
ibis 鹮 74
iguana 鬣蜥 34
inland taipan 内陆太攀蛇 81
insects 昆虫 24-25
internal organs 内脏 15
ivory gull 白鸥 98
jackrabbit 杰克兔 54
jaguar 美洲豹 30
jaw 颌 18, 19
jellyfish 水母 23
jerboa 跳鼠 81
jewel beetle 彩虹吉丁虫 30
jointed leg 节状腿 25
kangaroo 袋鼠 62
kangaroo rat 更格卢鼠 80
keel-billed toucan 厚嘴鹀鹀 34
kestrel 红隼 61
kingfisher 翠鸟 70
kite 鸢 50
koala 树袋熊 63
koi 锦鲤 73
lake 湖泊 66-67, 72-73
lantern fish 灯笼鱼 95
larva 幼虫 25
leafcutter ant 切叶蚁 31
leg 腿 15, 20, 24
lemming 旅鼠 102
lemur 狐猴 32
leopard 豹 28
Life in the ocean 海洋动物 82-95
life cycle of a butterfly 蝴蝶的生命周期 25
life cycle of a frog 蛙的生命周期 21
limpet 帽贝 86
lion 狮 56

liver 肝 15
lizard 蜥蜴 19, 28, 33, 34, 55, 62, 73, 78
llama 家羊驼 51
lobster 龙虾 22
locust 蝗虫 80
lower slopes 山麓 48-49
lowland gorilla 低地大猩猩 31
lungfish 肺鱼 74
lung 肺 15
lynx 猞猁 44
macaw 金刚鹦鹉 34
mackerel 鲭 88
major rivers 主要河流 67
mallard 绿头鸭 70
mammal 哺乳动物 14-15
manatee 海牛 90
mane 鬃毛 14
manta ray 蝠鲼 90
marlin 马林鱼 89
marmoset 狨 34
marsupial 有袋类动物 15
mayfly 蜉蝣 70
meerkat 细尾獴 58
millipede 马陆 42
minibeasts 小型无脊椎动物 24-25, 42, 70, 104
mink 水貂 68
minnow 鲦 73
moist skin 湿润的表皮 20
mole 鼹鼠 42
mongoose 獴 58
monitor lizard 巨蜥 78
monkey 猴子 15, 32, 33, 34
monkfish 鮟鱇 89
moorhen 黑水鸡 70
moorland 高沼地 55, 60-61
moose 驼鹿 45
morpho butterfly 闪蝶 28, 34

120

English-Chinese index 英汉索引

moth 蛾 60
Mountain animals 山地动物 46-51
mountain birds 山地鸟类 50
mountain goat 雪羊 48
mountain gorilla 山地大猩猩 49
mountain ranges 山脉 49
mouth 嘴；口器 20, 23, 91
mudskipper 弹涂鱼 84
musk ox 麝牛 103
muskrat 麝鼠 66
mussel 贻贝 86
muzzle 口鼻 14
neck 颈 17
neon tetra 霓虹脂鲤 68
newt 水栖蝾螈 73
nightingale 夜莺 40
nostril 鼻孔 14, 18
ocean 大洋 82-95
octopus 章鱼 23
okapi 㺢㹢狓 31
open ocean 外海 85
oral arm 口腕 23
orang-utan 猩猩 32
orca 虎鲸 91
osprey 鹗 73
ostrich 鸵鸟 54
otter 水獭 70
owl 猫头鹰；鸮 39, 41, 45, 103
ox 牛 103
oystercatcher 蛎鹬 86
panda 熊猫 51
panther 美洲狮 75
parrot 鹦鹉 17, 34
parrotfish 鹦嘴鱼 92
partridge 山鹑 61
paw 脚掌 14
pectoral fin 胸鳍 22
pelican 鹈鹕 72
pelvic fin 腹鳍 22
penguin 企鹅 98, 100, 101

peregrine falcon 游隼 50
pheasant 环颈雉 55
pigeon 鸽子 41
pike 狗鱼 68
pilot fish 舟䲟 88
pincer claw 螯钳 22
pine marten 松貂 39
piranha 食人鲳 69
plaice 鲽 88
platypus 鸭嘴兽 69
poison dart frog 箭毒蛙 33, 34
polar bear 北极熊 98
Polar regions 极地 96-103
polecat 鸡鼬 45
pollen basket 花粉篮 25
pond 池塘 66, 72-73
pond skater 水黾 70
porcupine 豪猪 45
porpoise 鼠海豚 90
pouch 育儿袋 15
prairie dog 草原犬鼠 61
prawn 明虾 84
praying mantis 螳螂 32
proboscis 口器 25
protruding eye 突出的眼睛 20
puffer fish 刺鲀 92
puffin 海鹦 86
pygmy marmoset 倭狨 34
python 蟒 58
queen 蜂王；蚁后 25, 59
raccoon 浣熊 40
rainbow lorikeet 彩虹吸蜜鹦鹉 32
rainbow trout 虹鳟 66
Rainforest creatures 雨林动物 26-35
rainforest floor 雨林林地层 30-31
rainforest layers 雨林层级 28
ray 鳐 90
red deer 欧洲马鹿 40
red fox 赤狐 41

red grouse 柳雷鸟 60
red kite 赤鸢 50
red panda 小熊猫 48
red squirrel 赤松鼠 39
red-eyed tree frog 红眼树蛙 28
reef 礁 84, 92-93
reptile 爬行动物 18-19
rhinoceros 犀牛 58
river 河流 66-71
River, lake and wetland wildlife 河流、湖泊和湿地的动物 64-76
roadrunner 走鹃 81
royal python 球蟒 58
salamander 蝾螈 20
sardine 沙丁鱼 88
savannah 稀树草原；萨瓦纳草原 56-59
scaly skin 鳞状皮肤 18
scarlet macaw 绯红金刚鹦鹉 34
scorpion 蝎子 78
sea 海 82-95
sea anemone 海葵 86
sea bass 海鲈 89
sea creatures 海洋动物 22-23, 82-95
sea cucumber 海参 84
sea goldie 丝鳍拟花鮨 93
sea lion 海狮 84
sea pen 海笔 93
sea urchin 海胆 86
seagull 海鸥 86
seahorse 海马 92
seal 海豹 98
seashore 海滩 84, 86-87
shallow seas 浅海 84
shark 鲨鱼 90
shell 甲壳 18
shrew 鼩鼱 42
shrimp 小虾 86
sidewinder rattlesnake 角响尾蛇 79
siphon 虹吸管 23

121

English-Chinese index 英汉索引

skink 石龙子　62
skunk 臭鼬　44
skylark 云雀　61
sloth 树懒　34
snail 蜗牛　33, 67
snake 蛇　19, 29, 31, 32, 55, 58, 68, 79
snapping turtle 蛇鳄龟　72
snout 吻　18
snow goose 雪雁　102
snow leopard 雪豹　49
snowy owl 雪鸮　103
sole 鳎　88
spectacled caiman 眼镜凯门鳄　67
sperm whale 抹香鲸　91
spider 蜘蛛　24, 28, 30, 31, 81
spider monkey 蜘蛛猴　34
spiny anteater 针鼹　62
sponge 海绵　94
spoonbill 琵鹭　67
squirrel 松鼠　39, 41
starfish 海星　86
Steller's jay 暗冠蓝鸦　44
stinger 螫针　25
stoat 白鼬　42
stomach 胃　15
stork 白鹳　75
sucker 吸盘　23
sugar glider 蜜袋鼯　29
sunbird 太阳鸟　29
swamp 木本沼泽　75
swan 天鹅　70
swimmeret 桡肢　22
swordfish 剑鱼　89
tadpole 蝌蚪　21
tail 尾巴；尾羽　14, 16, 20
tail fan 尾足　22
taipan 太攀蛇　81
talon 爪　16
tapir 貘　30
tarantula 狼蛛　28

tarsier 眼镜猴　33
Tasmanian devil 袋獾　41
tawny owl 灰林鸮　41
teat 乳头　15
teeth 牙齿　14
temperate rainforest 温带雨林　29
tentacle 触手　23
termite 白蚁　59
termite mound 白蚁丘　59
terrapin 水龟　75
thorax 胸部　24
three-toed sloth 三趾树懒　34
thumb 拇指　15
tiger 老虎　75
toad 蟾蜍　72
toe 脚趾　15, 20
tongue 舌　14
tortoise 陆龟　19, 79
toucan 鵎鵼　34
tree frog 树蛙　28
tree kangaroo 树袋鼠　32
tree snail 树蜗牛　33
treecreeper 旋木雀　45
triggerfish 鳞鲀　93
tripod fish 短吻三刺鲀　95
tropical rainforest 热带雨林　29
trout 鳟　66
tuna 金枪鱼　89
tundra 苔原　99
turkey 火鸡　40
turkey vulture 红头美洲鹫　81
turtle 海龟　18, 72, 93
underbelly 下腹部　18
understory 林下层　28
upper slopes 山顶斜坡　48-49
urchin 海胆　86
vampire squid 幽灵蛸　85
venom 毒液　19
venom sac 毒液囊　19
vent 肛门　22

vent fish 暖绵鳚　94
vicuña 骆马　48
vine snake 藤蛇　31
viper 蝰蛇　19, 32, 81
viperfish 蝰鱼　95
vulture 鹫　48, 81
walking leg 步行足　22
wallaby 沙袋鼠　62
warning colours 警戒色　33
warthog 疣猪　59
water boatman 松藻虫　70
water shrew 水鼩鼱　70
water snail 田螺　67
water vole 水䶄　70
weasel 伶鼬　41
webbed feet 蹼足　17
Weddell seal 韦德尔海豹　98
weevil 象甲　33
wetland 湿地　67, 74-75
whale 鲸　91, 98
wide mouth 宽大的嘴巴　20
wild turkey 野火鸡　40
wildebeest 角马　59
wing 翅膀　16, 24
wing case 翅鞘　24
wingtip 翅尖　16
wolf 狼　39, 99
wolf fish 狼鳚　85
wolverine 狼獾　45
wombat 毛鼻袋熊　55
wood pigeon 林鸽　41
wood warbler 林莺　41
woodland floor 林地地面　42-43
woodlouse 鼠妇　42
woodpecker 啄木鸟　38
woolly monkey 绒毛猴　34
worker 工蜂　25
worm 蠕虫　42
yak 牦牛　51
zebra 斑马　56

122

Chinese-English index 汉英索引

A

阿氏沙龟 (ā shì shā guī) desert tortoise 79
鮟鱇 (ān kāng) monkfish 89
暗冠蓝鸦 (àn guān lán yā) Steller's jay 44
螯 (áo) claw 22
螯钳 (áo qián) pincer claw 22
澳大利亚荒野 (ào dà lì yà huāng yě) Australian bush 55, 62-63
澳洲斧螳 (ào zhōu fǔ táng) giant rainforest praying mantis 32
澳洲巨蜥 (ào zhōu jù xī) goanna 55
澳洲野犬 (ào zhōu yě quǎn) dingo 62

B

白鹳 (bái guàn) stork 75
白鲸 (bái jīng) beluga whale 98
白鸥 (bái ōu) ivory gull 98
白蚁 (bái yǐ) termite 59
白蚁丘 (bái yǐ qiū) termite mound 59
白鼬 (bái yòu) ermine 99
白鼬 (bái yòu) stoat 42
斑马 (bān mǎ) zebra 56
宝贝 (bǎo bèi) cowry 92
保护色 (bǎo hù sè) camouflage 19
豹 (bào) leopard 28
北极地区 (běi jí dì qū) Arctic 98-101
北极狐 (běi jí hú) Arctic fox 103
北极狼 (běi jí láng) Arctic wolf 99
北极旅鼠 (běi jí lǚ shǔ) Arctic lemming 102
北极兔 (běi jí tù) Arctic hare 102
北极熊 (běi jí xióng) polar bear 98
背部 (bèi bù) back 14
背鳍 (bèi qí) dorsal fin 22
鼻孔 (bí kǒng) nostril 14, 18
壁虎 (bì hǔ) gecko 33
臂 (bì) arm 15
捕鸟蛛 (bǔ niǎo zhū) goliath spider 31
哺乳动物 (bǔ rǔ dòng wù) mammal 15-15
步行足 (bù xíng zú) walking leg 22

C

彩虹吉丁虫 (cǎi hóng jí dīng chóng) jewel beetle 30
彩虹吸蜜鹦鹉 (cǎi hóng xī mì yīng wǔ) rainbow lorikeet 32
草原动物 (cǎo yuán dòng wù) Grassland wildlife 52-63
草原犬鼠 (cǎo yuán quǎn shǔ) prairie dog 61
叉状舌 (chā zhuàng shé) forked tongue 19
蟾蜍 (chán chú) toad 72
长鼻浣熊 (cháng bí huàn xióng) coatimundi 34
长臂猿 (cháng bì yuán) gibbon 29
长颈鹿 (cháng jǐng lù) giraffe 57
长颈象甲 (cháng jǐng xiàng jiǎ) giraffe weevil 33
长鬣蜥 (cháng liè xī) Chinese water dragon lizard 19
常绿林 (cháng lǜ lín) evergreen forest 39
池塘 (chí táng) pond 66, 72-73
赤狐 (chì hú) red fox 41
赤松鼠 (chì sōng shǔ) red squirrel 39
赤鸢 (chì yuān) red kite 50
翅尖 (chì jiān) wingtip 16
翅膀 (chì bǎng) wing 16, 24
翅鞘 (chì qiào) wing case 24
虫 (chóng) bug 30
臭鼬 (chòu yòu) skunk 44
触角 (chù jiǎo) antenna 22, 24
触手 (chù shǒu) tentacle 23
刺鲀 (cì tún) puffer fish 92
刺猬 (cì wei) hedgehog 42
翠鸟 (cuì niǎo) kingfisher 70

D

大舒 (dà yú) barracuda 88
大比目鱼 (dà bǐ mù yú) halibut 88
大砗磲 (dà chē qú) giant clam 92
大食蚁兽 (dà shí yǐ shòu) giant anteater 28
大王乌贼 (dà wáng wū zéi) giant squid 95
大猩猩 (dà xīng xing) gorilla 31, 49
大型猫科动物 (dà xíng māo kē dòng wù) big cat 28, 30, 39, 44, 50, 56, 75
大熊猫 (dà xióng māo) giant panda 51
大洋 (dà yáng) ocean 82-95
袋獾 (dài huān) Tasmanian devil 41

123

Chinese-English index 汉英索引

袋狸 (dài lí) bandicoot 63
袋鼠 (dài shǔ) kangaroo 62
单峰驼 (dān fēng tuó) dromedary camel 80
淡水鳌虾 (dàn shuǐ áo xiā) crayfish 68
灯笼鱼 (dēng long yú) lantern fish 95
瞪羚 (dèng líng) gazelle 54
低地大猩猩 (dī dì dà xīng xing) lowland gorilla 31
帝企鹅 (dì qǐ é) emperor penguin 98
帝王蛾 (dì wáng é) emperor moth 60
电鳗 (diàn mán) electric eel 69
雕 (diāo) eagle 16, 29, 50
鲽 (dié) plaice 88
豆娘 (dòu niáng) damselfly 66
毒液 (dú yè) venom 19
毒液囊 (dú yè náng) venom sac 19
短尾猫 (duǎn wěi māo) bobcat 39
短吻鳄 (duǎn wěn è) alligator 74
短吻三刺鲀 (duǎn wěn sān cì tún) tripod fish 95

E

鹅 (é) goose 17
蛾 (é) moth 60
鹗 (è) osprey 73
鳄鱼 (è yú) crocodile 18, 68
鸸鹋 (ér miáo) emu 63
耳朵 (ěr duo) ear 14

F

飞蜥 (fēi xī) flying dragon lizard 28
飞鱼 (fēi yú) flying fish 85
非洲森林象 (fēi zhōu sēn lín xiàng) African forest elephant 31
非洲象 (fēi zhōu xiàng) African elephant 57
绯红金刚鹦鹉 (fēi hóng jīn gāng yīng wǔ) scarlet macaw 34
鲱 (fēi) herring 85
肺 (fèi) lung 15
肺鱼 (fèi yú) lungfish 74
粉红凤头鹦鹉 (fěn hóng fèng tóu yīng wǔ) galah 62
蜂鸟 (fēng niǎo) hummingbird 33
蜂王 (fēng wáng) queen 25
蜂窝 (fēng wō) beehive 25

凤头鹦鹉 (fèng tóu yīng wǔ) cockatoo 63
佛罗里达美洲狮 (fó luó lǐ dá měi zhōu shī) Florida panther 75
蜉蝣 (fú yóu) mayfly 70
蝠鲼 (fú fèn) manta ray 90
腹部 (fù bù) abdomen 22, 24
腹鳍 (fù qí) pelvic fin 22

G

肝 (gān) liver 15
肛门 (gāng mén) vent 22
高沼地 (gāo zhǎo dì) moorland 55, 60-61
鸽子 (gē zi) pigeon 41
更格卢鼠 (gēng gé lú shǔ) kangaroo rat 80
工蜂 (gōng fēng) worker 25
狗 (gǒu) dog 14
狗鲨 (gǒu shā) dogfish 88
狗鱼 (gǒu yú) pike 68
骨顶鸡 (gǔ dǐng jī) coot 66
鼓膜 (gǔ mó) eardrum 20
灌木 (guàn mù) bush 62
聊狐 (guō hú) fennec fox 78

H

海 (hǎi) sea 82-95
海豹 (hǎi bào) seal 98
海笔 (hǎi bǐ) sea pen 93
海参 (hǎi shēn) sea cucumber 84
海胆 (hǎi dǎn) sea urchin 86
海胆 (hǎi dǎn) urchin 86
海龟 (hǎi guī) turtle 18, 72, 93
海葵 (hǎi kuí) anemone 86
海葵 (hǎi kuí) sea anemone 86
海鲈 (hǎi lú) sea bass 89
海马 (hǎi mǎ) seahorse 92
海绵 (hǎi mián) sponge 94
海牛 (hǎi niú) manatee 90
海鸥 (hǎi ōu) seagull 86
海狮 (hǎi shī) sea lion 84
海滩 (hǎi tān) seashore 84, 86-87
海豚 (hǎi tún) dolphin 90
海星 (hǎi xīng) starfish 86

Chinese-English index 汉英索引

海洋动物 (hǎi yáng dòng wù) Life in the ocean 82-95
海洋动物 (hǎi yáng dòng wù) sea creatures 22-23, 82-95
海鹦 (hǎi yīng) puffin 86
豪猪 (háo zhū) porcupine 45
河狸 (hé lí) beaver 69
河狸巢穴 (hé lí cháo xué) beaver lodge 69
河狸鼠 (hé lí shǔ) coypu 69
河流 (hé liú) river 66-71
河流、湖泊和湿地的动物 (hé liú、hú pō hé shī dì de dòng wù) River, lake and wetland wiidlife 64-75
河马 (hé mǎ) hippopotamus 66
河乌 (hé wū) dipper 70
颌 (hé) jaw 18-19
赫氏陆龟 (hè shì lù guī) Hermann's tortoise 19
鹤 (hè) crane 74
鹤鸵 (hè tuó) cassowary 31
黑斑蛙 (hēi bān wā) common frog 70
黑寡妇蜘蛛 (hēi guǎ fu zhī zhū) black widow spider 81
黑水鸡 (hēi shuǐ jī) moorhen 70
黑线鳕 (hēi xiàn xuě) haddock 89
黑猩猩 (hēi xīng xing) chimpanzee 29
黑熊 (hēi xióng) black bear 44
红鹳 (hóng guàn) flamingo 72
红玫瑰鹦鹉 (hóng méi gui yīng wǔ) crimson rosella 49
红隼 (hóng sǔn) kestrel 61
红头美洲鹫 (hóng tóu měi zhōu jiù) turkey vulture 81
红眼树蛙 (hóng yǎn shù wā) red-eyed tree frog 28
虹吸管 (hóng xī guǎn) siphon 23
虹鳟 (hóng zūn) rainbow trout 66
猴子 (hóu zi) monkey 15, 32, 33, 34
吼猴 (hǒu hóu) howler monkey 34
后翅 (hòu chì) hindwing 25
后腿 (hòu tuǐ) hind leg 14
厚嘴鹲鹈 (hòu zuǐ tuǒ kōng) keel-billed toucan 34
狐猴 (hú hóu) lemur 32
狐狸 (hú li) fox 41, 78, 103

胡兀鹫 (hú wù jiù) bearded vulture 48
湖泊 (hú pō) lake 66-67, 72-73
蝴蝶 (hú dié) butterfly 25, 28, 38
蝴蝶的生命周期 (hú dié de shēng mìng zhōu qī) life cycle of a butterfly 25
蝴蝶鱼 (hú dié yú) butterfly fish 93
虎鲸 (hǔ jīng) orca 91
虎皮鹦鹉 (hǔ pí yīng wǔ) budgerigar 63
花粉篮 (huā fěn lán) pollen basket 25
花鼠 (huā shǔ) chipmunk 38
花园蜗牛 (huā yuán wō niú) garden snail 38
獾 (huān) badger 42
环颈雉 (huán jǐng zhì) pheasant 55
鹮 (huán) ibis 74
浣熊 (huàn xióng) raccoon 40
荒漠动物 (huāng mò dòng wù) Desert wildlife 76-81
黄尾副刺尾鱼 (huáng wěi fù cì wěi yú) blue tang 84
蝗虫 (huáng chóng) locust 80
灰林鸮 (huī lín xiāo) tawny owl 41
灰松鼠 (huī sōng shǔ) grey squirrel 41
灰熊 (huī xióng) grizzly bear 103
喙 (huì) bill 17
喙 (huì) beak 16, 23
火鸡 (huǒ jī) turkey 40
獾狐狓 (huò jiā pí) okapi 31

J

鸡鼬 (jī yòu) polecat 45
极北蝰 (jí běi kuí) common adder 55
极地 (jí dì) Polar regions 96-103
加拿大雁 (jiā ná dà yàn) Canada goose 99
家羊驼 (jiā yáng tuó) llama 51
甲虫 (jiǎ chóng) beetle 24, 29, 30, 33
甲壳 (jiǎ qiào) carapace 22
甲壳 (jiǎ qiào) shell 18
剑鱼 (jiàn yú) swordfish 89
箭毒蛙 (jiàn dú wā) poison dart frog 33, 34
郊狼 (jiāo láng) coyote 60
礁 (jiāo) reef 84, 92-93
角蝰 (jiǎo kuí) horned desert viper 81
角马 (jiǎo mǎ) wildebeest 59

125

Chinese-English index 汉英索引

角响尾蛇 (jiǎo xiǎng wěi shé) sidewinder rattlesnake 79
脚掌 (jiǎo zhǎng) paw 14
脚趾 (jiǎo zhǐ) toe 15, 20
节状腿 (jié zhuàng tuǐ) jointed leg 25
杰克兔 (jié kè tù) jackrabbit 54
结肠 (jié cháng) colon 15
睫角棕榈蝮 (jié jiǎo zōng lǘ fù) eyelash viper 32
金雕 (jīn diāo) golden eagle 50
金刚鹦鹉 (jīn gāng yīng wǔ) macaw 34
金枪鱼 (jīn qiāng yú) tuna 89
金鱼 (jīn yú) goldfish 73
锦鲤 (jǐn lǐ) koi 73
鲸 (jīng) whale 91, 98
鲸须 (jīng xū) baleen 91
颈 (jǐng) neck 17
警戒色 (jǐng jiè sè) warning colours 33
鹫 (jiù) vulture 48, 81
巨管虫 (jù guǎn chóng) giant tube worm 94
巨花潜金龟 (jù huā qián jīn guī) goliath beetle 29
巨犀金龟 (jù xī jīn guī) Hercules beetle 30
巨蜥 (jù xī) monitor lizard 78
卷尾猴 (juǎn wěi hóu) capuchin monkey 32

K

凯门鳄 (kǎi mén è) caiman 67
蝌蚪 (kē dǒu) tadpole 21
口鼻 (kǒu bí) muzzle 14
口器 (kǒu qì) proboscis 25
口器 (kǒu qì) mouth 23
口腕 (kǒu wàn) oral arm 23
宽大的嘴巴 (kuān dà de zuǐ ba) wide mouth 20
鵟 (kuáng) buzzard 50
蝰 (kuí) adder 55
蝰蛇 (kuí shé) viper 19, 32, 81
蝰鱼 (kuí yú) viperfish 94
昆虫 (kūn chóng) insects 24-25

L

蓝鲸 (lán jīng) blue whale 91
蓝舌蜥 (lán shé xī) blue-tongued skink 62
狼 (láng) wolf 39, 99
狼獾 (láng huān) wolverine 45
狼鳚 (láng wèi) wolf fish 85
狼蛛 (láng zhū) tarantula 28
老虎 (lǎo hǔ) tiger 75
类人猿 (lèi rén yuán) ape 15, 29, 31, 32, 49
冷荒漠 (lěng huāng mò) cold desert 79
鲤鱼 (lǐ yú) carp 73
蛎鹬 (lì yù) oystercatcher 86
两栖动物 (liǎng qī dòng wù) amphibian 20
猎豹 (liè bào) cheetah 57
猎蝽 (liè chūn) assassin bug 30
鬣蜥 (liè xī) iguana 34
林地层 (lín dì céng) forest floor 28
林地地面 (lín dì dì miàn) woodland floor 42-43
林鸽 (lín gē) wood pigeon 41
林上层 (lín shàng céng) emergent layer 28
林下层 (lín xià céng) understory 28
林莺 (lín yīng) wood warbler 41
鳞鲀 (lín tún) triggerfish 93
鳞状皮肤 (lín zhuàng pí fū) scaly skin 18
伶鼬 (líng yòu) weasel 41
羚羊 (líng yáng) antelope 57
硫磺蝶 (liú huáng dié) brimstone butterfly 38
柳雷鸟 (liǔ léi niǎo) red grouse 60
龙虾 (lóng xiā) lobster 22
漏斗网蛛 (lòu dǒu wǎng zhū) funnel web spider 30
鸬鹚 (lú cí) cormorant 86
陆龟 (lù guī) tortoise 19, 79
鹿 (lù) deer 38, 40
鹭 (lù) heron 17, 70
卵 (luǎn) egg 16, 25, 59, 69
骆驼 (luò tuo) camel 79, 80
骆马 (luò mǎ) vicuña 48
落叶林 (luò yè lín) deciduous forest 39-41
旅鼠 (lǚ shǔ) lemming 102
绿海龟 (lǜ hǎi guī) green turtle 93
绿树蚺 (lǜ shù rán) emerald tree boa 29
绿头鸭 (lǜ tóu yā) mallard 70

M

马 (mǎ) horse 14

126

Chinese-English index 汉英索引

马林鱼 (mǎ lín yú) marlin　　89
马陆 (mǎ lù) millipede　　42
蚂蚁 (mǎ yǐ) ant　　31, 80
蚂蚱 (mà zha) grasshopper　　61
鳗 (mán) eel　　69, 95
盲鳗 (máng mán) hagfish　　95
蟒 (mǎng) python　　58
猫头鹰 (māo tóu yīng) owl　　39, 41, 45, 103
毛鼻袋熊 (máo bí dài xióng) wombat　　55
毛虫 (máo chóng) caterpillar　　25
毛丝鼠 (máo sī shǔ) chinchilla　　49
牦牛 (máo niú) yak　　51
帽贝 (mào bèi) limpet　　86
美洲豹 (měi zhōu bào) jaguar　　30
美洲角雕 (měi zhōu jiǎo diāo) harpy eagle　　29
美洲鬣蜥 (měi zhōu liè xī) green iguana　　34
美洲牛蛙 (měi zhōu niú wā) American bullfrog　　72
美洲狮 (měi zhōu shī) cougar　　50
美洲狮 (měi zhōu shī) panther　　75
獴 (měng) mongoose　　58
孟加拉虎 (mèng jiā lā hǔ) Bengal tiger　　75
蜜袋鼯 (mì dài wú) sugar glider　　29
蜜蜂 (mì fēng) bee　　25
蜜蜂 (mì fēng) honey bee　　25
蜜壶蚁 (mì hú yǐ) honeypot ant　　80
缅甸蟒 (miǎn diàn mǎng) Burmese python　　19
明虾 (míng xiā) prawn　　84
抹香鲸 (mǒ xiāng jīng) sperm whale　　91
貘 (mò) tapir　　30
拇指 (mǔ zhǐ) thumb　　15
木本沼泽 (mù běn zhǎo zé) swamp　　75

N

南极地区 (nán jí dì qū) Antarctic　　98
南极洲 (nán jí zhōu) Antarctica　　98
脑 (nǎo) brain　　15
内陆太攀蛇 (nèi lù tài pān shé) inland taipan　　81
内脏 (nèi zàng) internal organs　　15
霓虹脂鲤 (ní hóng zhī lǐ) neon tetra　　68
鲇鱼 (nián yú) catfish　　66
鸟 (niǎo) bird　　16-17

牛 (niú) ox　　103
牛蛙 (niú wā) bullfrog　　72
暖绵鳚 (nuǎn mián wèi) vent fish　　94

O

欧洲马鹿 (ōu zhōu mǎ lù) red deer　　40

P

爬行动物 (pá xíng dòng wù) reptile　　18-19
膀胱 (páng guāng) bladder　　15
螃蟹 (páng xiè) crab　　86, 95
皮褶 (pí zhě) hood　　19
琵鹭 (pí lù) spoonbill　　67
偏口鱼 (piān kǒu yú) flounder　　88
破碎螯 (pò suì áo) crusher claw　　22
普通鵟 (pǔ tōng kuáng) common buzzard　　50
普通蟾蜍 (pǔ tōng chán chú) common toad　　72
普通鲤鱼 (pǔ tōng lǐ yú) common carp　　73
蹼足 (pǔ zú) webbed feet　　17

Q

鳍 (qí) fin　　22
鳍足 (qí zú) flipper　　18
企鹅 (qǐ é) penguin　　98, 100, 101
前翅 (qián chì) forewing　　25
前腿 (qián tuǐ) foreleg　　14
浅海 (qiǎn hǎi) shallow seas　　84
切叶蚁 (qiē yè yǐ) leafcutter ant　　31
蜻蜓 (qīng tíng) dragonfly　　70
鲭 (qīng) mackerel　　88
蚯蚓 (qiū yǐn) earthworm　　42
犰狳 (qiú yú) armadillo　　59
球蟒 (qiú mǎng) royal python　　58
鼩鼱 (qú jīng) shrew　　42
雀鲷 (què diāo) damselfish　　93

R

桡肢 (ráo zhī) swimmeret　　22
热带雨林 (rè dài yǔ lín) tropical rainforest　　29
热荒漠 (rè huāng mò) hot desert　　78
狨 (róng) marmoset　　34
绒毛猴 (róng máo hóu) woolly monkey　　34
绒鸭 (róng yā) eider duck　　102

127

Chinese-English index 汉英索引

蝾螈 (róng yuán) salamander　20
肉垂 (ròu chuí) dewlap　20
儒艮 (rú gèn) dugong　90
蠕虫 (rú chóng) worm　42
乳头 (rǔ tóu) teat　15

S

萨瓦纳草原 (sà wǎ nà cǎo yuán) savannah　56-59
鳃 (sāi) gill　21, 23
三趾树懒 (sān zhǐ shù lǎn) three-toed sloth　34
伞状体 (sǎn zhuàng tǐ) bell　23
森林和林地动物 (sēn lín hé lín dì dòng wù) Forest and woodland wildlife　36-45
沙袋鼠 (shā dài shǔ) wallaby　62
沙丁鱼 (shā dīng yú) sardine　88
沙鼠 (shā shǔ) gerbil　80
鲨鱼 (shā yú) shark　90
山鹑 (shān chún) partridge　61
山地大猩猩 (shān dì dà xīng xing) mountain gorilla　49
山地动物 (shān dì dòng wù) Mountain animals　46-51
山地鸟类 (shān dì niǎo lèi) mountain birds　50
山顶斜坡 (shān dǐng xié pō) upper slopes　48-49
山麓 (shān lù) lower slopes　48-49
山脉 (shān mài) mountain ranges　49
山羊 (shān yáng) goat　48
珊瑚礁 (shān hú jiāo) coral reef　84, 92-93
珊瑚蛇 (shān hú shé) coral snake　29
闪蝶 (shǎn dié) morpho butterfly　28, 34
猞猁 (shē lì) lynx　44
舌 (shé) tongue　14
蛇 (shé) snake　19, 29, 31, 32, 55, 58, 68, 79
蛇鳄龟 (shé è guī) snapping turtle　72
麝牛 (shè niú) musk ox　103
麝鼠 (shè shǔ) muskrat　66
深海鮟鱇 (shēn hǎi ān kāng) angler fish　94
深海动物 (shēn hǎi dòng wù) deep-sea creatures　94
深海海绵 (shēn hǎi hǎi mián) deep-sea sponge　94
深海蜘蛛蟹 (shēn hǎi zhī zhū xiè) deep-sea spider crab　95

神仙鱼 (shén xiān yú) angelfish　92
狮 (shī) lion　56
湿地 (shī dì) wetland　67, 74-75
湿润的表皮 (shī rùn de biǎo pí) moist skin　20
石龙子 (shí lóng zǐ) skink　62
食人鲳 (shí rén chāng) piranha　69
食蚁兽 (shí yǐ shòu) anteater　28
螫针 (shì zhēn) stinger　25
手指 (shǒu zhǐ) finger　15
鼠妇 (shǔ fù) woodlouse　42
鼠海豚 (shǔ hǎi tún) porpoise　90
树袋鼠 (shù dài shǔ) tree kangaroo　32
树袋熊 (shù dài xióng) koala　63
树冠层 (shù guān céng) canopy　28, 32-33
树懒 (shù lǎn) sloth　34
树蛙 (shù wā) tree frog　28
树蜗牛 (shù wō niú) tree snail　33
双峰驼 (shuāng fēng tuó) Bactrian camel　79
双嵴冠蜥 (shuāng jí guàn xī) basilisk lizard　73
双髻鲨 (shuāng jì shā) hammerhead shark　90
双角犀鸟 (shuāng jiǎo xī niǎo) great hornbill　28
水貂 (shuǐ diāo) mink　68
水龟 (shuǐ guī) terrapin　75
水黾 (shuǐ mǐn) pond skater　70
水母 (shuǐ mǔ) jellyfish　23
水牛 (shuǐ niú) buffalo　56
水䶄 (shuǐ píng) water vole　70
水栖蝾螈 (shuǐ qī róng yuán) newt　73
水鼩鼱 (shuǐ qú jīng) water shrew　70
水蚺 (shuǐ rán) anaconda　68
水獭 (shuǐ tǎ) otter　70
水豚 (shuǐ tún) capybara　72
睡鼠 (shuì shǔ) dormouse　42
丝鳍拟花鮨 (sī qí nǐ huā yì) sea goldie　93
松貂 (sōng diāo) pine marten　39
松鸡 (sōng jī) grouse　60
松鼠 (sōng shǔ) squirrel　39, 41
松藻虫 (sōng zǎo chóng) greater water boatman　70
松藻虫 (sōng zǎo chóng) water boatman　70
隼 (sǔn) falcon　50

128

Chinese-English index 汉英索引

T

鳎 (tǎ) sole   88
苔原 (tái yuán) tundra   99
太攀蛇 (tài pān shé) taipan   81
太阳鸟 (tài yáng niǎo) sunbird   29
弹涂鱼 (tán tú yú) mudskipper   84
螳螂 (táng láng) praying mantis   32
藤壶 (téng hú) barnacle   86
藤蛇 (téng shé) vine snake   31
鹈鹕 (tí hú) pelican   72
蹄 (tí) hoof   14
天鹅 (tiān é) swan   70
黇鹿 (tiān lù) fallow deer   38
田螺 (tián luó) water snail   67
鲦 (tiáo) minnow   73
跳鼠 (tiào shǔ) jerboa   81
头部 (tóu bù) head   23, 24
突出的眼睛 (tū chū de yǎn jing) protruding eye   20
土豚 (tǔ tún) aardvark   58
兔耳袋狸 (tù ěr dài lí) bilby   63
腿 (tuǐ) leg   15, 19, 24
吞噬鳗 (tūn shì mán) gulper eel   60
豚鼠 (tún shǔ) cavy   51
驼鹿 (tuó lù) moose   45
鸵鸟 (tuó niǎo) ostrich   54
鵎鵼 (tuǒ kōng) toucan   34

W

蛙 (wā) frog   20, 21, 28, 33, 70, 72
蛙的生命周期 (wā de shēng mìng zhōu qī) life cycle of a frog   21
蛙卵 (wā luǎn) frogspawn   21
外海 (wài hǎi) open ocean   85
腕足 (wàn zú) arm   23
韦德尔海豹 (wéi dé ěr hǎi bào) Weddell seal   98
尾巴 (wěi ba) tail   14, 20
尾鳍 (wěi qí) caudal fin   22
尾羽 (wěi yǔ) tail   16
尾足 (wěi zú) tail fan   22
胃 (wèi) stomach   15
温带雨林 (wēn dài yǔ lín) temperate rainforest   29

吻 (wěn) snout   18
倭狨 (wō róng) pygmy marmoset   34
蜗牛 (wō niú) snail   33, 67
乌林鸮 (wū lín xiāo) great grey owl   45
乌贼 (wū zéi) cuttlefish   84
蜈蚣 (wú gōng) centipede   42
鼯猴 (wú hóu) colugo   33

X

吸蜜鸟 (xī mì niǎo) honey eater   32
吸盘 (xī pán) sucker   23
稀树草原 (xī shù cǎo yuán) savannah   56-59
犀鸟 (xī niǎo) hornbill   28
犀牛 (xī niú) rhinoceros   58
蜥蜴 (xī yì) lizard   19, 28, 33, 34, 55, 62, 73, 78
细尾獴 (xì wěi měng) meerkat   58
下腹部 (xià fù bù) underbelly   18
象 (xiàng) elephant   31, 57
象甲 (xiàng jiǎ) weevil   33
鸮 (xiāo) owl   39, 41, 45, 103
小丑鱼 (xiǎo chǒu yú) clown fish   93
小虾 (xiǎo xiā) shrimp   86
小型无脊椎动物 (xiǎo xíng wú jǐ zhuī dòng wù) minibeasts   24-25, 42, 70, 104
小熊猫 (xiǎo xióng māo) red panda   49
小羊驼 (xiǎo yáng tuó) alpaca   51
蝎子 (xiē zi) scorpion   78
胁腹 (xié fù) flank   14
心脏 (xīn zàng) heart   15
信天翁 (xìn tiān wēng) albatross   98
猩猩 (xīng xing) orang-utan   32
行军蚁 (xíng jūn yǐ) army ant   30
胸部 (xiōng bù) breast   17
胸部 (xiōng bù) chest   15
胸部 (xiōng bù) thorax   24
胸鳍 (xiōng qí) pectoral fin   22
雄蜂 (xióng fēng) drone   25
熊 (xióng) bear   44, 98, 103
熊猫 (xióng māo) panda   51
旋木雀 (xuán mù què) treecreeper   45
雪豹 (xuě bào) snow leopard   49

129

Chinese-English index 汉英索引

雪貂 (xuě diāo) ferret 61
雪鸮 (xuě xiāo) snowy owl 103
雪雁 (xuě yàn) snow goose 102
雪羊 (xuě yáng) mountain goat 48
鳕 (xuě) cod 89
驯化的动物 (xùn huà de dòng wù) domesticated creatures 51
驯鹿 (xùn lù) caribou 102

Y

鸭 (yā) duck 17, 102
鸭嘴兽 (yā zuǐ shòu) platypus 69
牙齿 (yá chǐ) teeth 14
亚马孙鹦哥 (yà mǎ sūn yīng gē) Amazon parrot 34
亚马孙雨林 (yà mǎ sūn yǔ lín) Amazon rainforest 34-35
眼 (yǎn) eye 14, 23
眼镜猴 (yǎn jìng hóu) tarsier 33
眼镜凯门鳄 (yǎn jìng kǎi mén è) spectacled caiman 67
鼹鼠 (yǎn shǔ) mole 42
雁 (yàn) goose 99
鳐 (yáo) ray 90
野火鸡 (yě huǒ jī) wild turkey 40
野牛 (yě niú) bison 54
野兔 (yě tù) hare 61, 102
夜莺 (yè yīng) nightingale 40
贻贝 (yí bèi) mussel 86
蚁后 (yǐ hòu) queen 59
臆羚 (yì líng) chamois 50
鹦鹉 (yīng wǔ) parrot 17, 34
鹦嘴鱼 (yīng zuǐ yú) parrotfish 92
鹰鸮 (yīng xiāo) hawk owl 39
蛹 (yǒng) chrysalis 25
幽灵蛸 (yōu líng shāo) vampire squid 85
疣猴 (yóu hóu) colobus monkey 33
疣猪 (yóu zhū) warthog 59
游隼 (yóu sǔn) peregrine falcon 50
有袋类动物 (yǒu dài lèi dòng wù) marsupial 15
幼虫 (yòu chóng) larva 25
幼崽 (yòu zǎi) baby 15

鱼类和其他海洋动物 (yú lèi hé qí tā hǎi yáng dòng wù) fish and other sea creatures 22-23, 82-95
羽毛 (yǔ máo) feather 16
雨林层级 (yǔ lín céng jí) rainforest layers 28
雨林动物 (yǔ lín dòng wù) **Rainforest creatures 26-35**
雨林林地层 (yǔ lín lín dì céng) rainforest floor 30-31
育儿袋 (yù ér dài) pouch 15
鸢 (yuān) kite 50
羱羊 (yuán yáng) ibex 51
云豹 (yún bào) clouded leopard 28
云雀 (yún què) skylark 61

Z

章鱼 (zhāng yú) octopus 23
爪 (zhǎo) talon 16
爪 (zhǎo) claw 14, 18
褶虎 (zhě hǔ) flying gecko 33
褶胸鱼 (zhě xiōng yú) hatchetfish 94
针鼹 (zhēn yǎn) spiny anteater 62
珍珠鸡 (zhēn zhū jī) guinea fowl 60
蜘蛛 (zhī zhū) spider 24, 28, 30, 31, 81
蜘蛛猴 (zhī zhū hóu) spider monkey 34
中空的毒牙 (zhōng kōng de dú yá) hollow fang 19
舟鲕 (zhōu shī) pilot fish 88
主要河流 (zhǔ yào hé liú) major rivers 67
啄木鸟 (zhuó mù niǎo) woodpecker 38
鬃毛 (zōng máo) mane 14
走鹃 (zǒu juān) roadrunner 81
足 (zú) foot 15
钻纹龟 (zuàn wén guī) diamondback terrapin 75
嘴 (zuǐ) mouth 20, 91
鳟 (zūn) trout 66
座头鲸 (zuò tóu jīng) humpback whale 91

Quiz answers 小测验答案

Animal hunters
动物捕食者

1. **assassin bug** 猎蝽 (p. 30)

2. **leech** 水蛭 (p. 9)

3. **electric eel** 电鳗 (p. 69)

Watch out!
当心！

4. **skunk** 臭鼬 (p. 44)

5. **puffer fish** 刺鲀 (p. 92)

6. **poison dart frog** 箭毒蛙 (p. 33)

Special features
特征

7. **armadillo** 犰狳 (p. 59)

8. **jerboa** 跳鼠 (p. 81)

9. **hummingbird** 蜂鸟 (p. 33)

Unusual behaviour
不寻常的行为

10. **basilisk lizard** 双嵴冠蜥 (p. 73)

Quiz answers 小测验答案

11. dolphin 海豚 (p. 90)

12. three-toed sloth
    三趾树懒 (p. 35)

Animal homes
动物的家

13. beaver 河狸 (p. 69)

14. termite 白蚁 (p. 59)

15. tapeworm 绦虫 (p. 9)

Animal parts
动物的身体部分

16. spider 蜘蛛 (p. 24)

17. frog 蛙 (p. 20)

18. octopus 章鱼 (p. 23)

What am I?
我是哪种动物？

19. platypus 鸭嘴兽 (p. 69)

20. lungfish 肺鱼 (p. 74)

图书在版编目（CIP）数据

牛津少儿英汉动物百科 / 英国牛津大学出版社编；张劲硕译. —北京：商务印书馆，2017
ISBN 978-7-100-12693-9

Ⅰ.①牛… Ⅱ.①英…②张… Ⅲ.①英语—少儿读物②动物—少儿读物 Ⅳ.①H319.4②Q95-49

中国版本图书馆CIP数据核字(2016)第262238号

所有权利保留。
未经许可，不得以任何方式使用。

牛津少儿英汉动物百科
Oxford English-Chinese
Visual Dictionary of Animals

张劲硕 译

商 务 印 书 馆 出 版
（北京王府井大街36号 邮政编码 100710）
商 务 印 书 馆 发 行
北京新华印刷有限公司印刷
ISBN 978-7-100-12693-9

2017年1月第1版　　开本 889×1194　1/16
2017年1月北京第1次印刷　印张 8½

定价：78.00元